BRITISH HIGHWAYS
AND BYWAYS FROM A MOTOR CAR

BRITISH HIGHWAYS
AND BYWAYS FROM A MOTOR CAR

THOMAS DOWLER MURPHY

ISBN: 978-93-5614-033-2

Published: 1908

LECTOR HOUSE LLP
E-MAIL: lectorpublishing@gmail.com

OLD HALF-TIMBERED HOUSES IN LEDBURY.

From Water Color by B. McGuinness.

BRITISH HIGHWAYS
AND BYWAYS FROM A MOTOR CAR

BEING A RECORD OF A FIVE THOUSAND MILE TOUR IN ENGLAND, WALES AND SCOTLAND

BY

THOMAS D. MURPHY

With Sixteen Illustrations in Colour and Thirty-two Duogravures
From Photographs; Also Two Descriptive Maps.

1908

A FOREWORD

In this chronicle of a summer's motoring in Britain I have not attempted a guide-book in any sense, yet the maps, together with the comments on highways, towns, and country, should be of some value even in that capacity. I hope, however, that the book, with its many illustrations and its record of visits to out-of-the way places, may be acceptable to those who may desire to tour Britain by rail or cycle as well as by motor car. Nor may it be entirely uninteresting to those who may not expect to visit the country in person but desire to learn more of it and its people. Although our journey did not follow the beaten paths of British touring, and while a motor car affords the most satisfactory means of reaching most of the places described, the great majority of these places are accessible by rail, supplemented in some cases by a walk or drive. A glance at the maps will indicate the large scope of country covered and the location of most places especially mentioned in the text.

It was not a tour of cities by any means, but of the most delightful country in the world, with its towns, villages, historic spots and solitary ruins. Whatever the merits or demerits of the text, there can be no question concerning the pictures. The color-plates were reproduced from original paintings by prominent artists, some of the pictures having been exhibited in the London Royal Academy. The thirty-two duogravures represent the very height of attainment in that process, being reproductions of the most perfect English photographs obtainable.

T.D.M.

January 1908.

FOREWORD TO SECOND EDITION

The first edition of BRITISH HIGHWAYS AND BYWAYS FROM A MOTOR CAR was printed from type—instead of from electrotype plates—thus giving an opportunity for additional care in the press work, with better results than with the ordinary book printed from plates. The publishers thought also that some time might elapse before a second edition would be called for. However, the unexpected happened and in less than a year a new edition is required.

This has afforded opportunity for numerous additions and corrections—since it was hardly possible that a book covering such a wide scope could be entirely free from mistakes, though, fortunately, these were mainly minor ones. I have to thank numerous readers for helpful suggestions.

That there is a distinct field for such a book is proven by the unexpectedly large demand for the first edition. I hope that the new and revised edition may meet with like favor.

T.D.M.

March 1, 1909.

CONTENTS

LIST OF ILLUSTRATIONS

LIST OF MAPS

OLD COTTAGE AT NORTON, NEAR EVESHAM.

From Water Color by G.F. Nicholls.

BRITISH HIGHWAYS
AND BYWAYS FROM A MOTOR CAR

I

A FEW GENERALITIES

Stratford-on-Avon stands first on the itinerary of nearly every American who proposes to visit the historic shrines of Old England. Its associations with Britain's immortal bard and with our own gentle Geoffrey Crayon are not unfamiliar to the veriest layman, and no fewer than thirty thousand pilgrims, largely from America, visit the delightful old town each year. And who ever came away disappointed? Who, if impervious to the charm of the place, ever dared to own it?

My first visit to Stratford-on-Avon was in the regulation fashion. Imprisoned in a dusty and comfortless first-class apartment—first-class is an irony in England when applied to railroad travel, a mere excuse for charging double—we shot around the curves, the glorious Warwickshire landscapes fleeting past in a haze or obscured at times by the drifting smoke. Our reveries were rudely interrupted by the shriek of the English locomotive—like an exaggerated toy whistle—and, with a mere glimpse of town and river, we were brought sharply up to the unattractive station of Stratford-on-Avon. We were hustled by an officious porter into an omnibus, which rattled through the streets until we landed at the Sign of the Red Horse; and the manner of our departure was even the same.

Just two years later, after an exhilarating drive of two or three hours over the broad, well-kept highway winding through the parklike fields, fresh from May showers, between Worcester and Stratford, our motor finally climbed a long hill, and there, stretched out before us, lay the valley of the Avon. Far away we caught the gleam of the immortal river, and rising from a group of splendid trees we beheld Trinity Church—almost unique in England for its graceful combination of massive tower and slender spire—the literary shrine of the English-speaking world, the enchanted spot where Shakespeare sleeps. About it were clustered the clean, tiled roofs of the charming town, set like a gem in the Warwickshire landscape, famous as the most beautiful section of Old England. Our car slowed to a stop, and only the subdued hum of the motor broke the stillness as we saw Stratford-on-Avon from afar, conscious of a beauty and sentiment that made our former visit seem commonplace indeed.

But I am not going to write of Stratford-on-Avon. Thousands have done this before me—some of them of immortal fame. I shall not attempt to describe or give details concerning a town that is probably visited each year by more people than any other place of the size in the world. I am simply striving in a few words to give the different impressions made upon the same party who visited the town twice in a comparatively short period, the first time by railway train and the last by motor car. If I have anything to say of Stratford, it will come in due sequence in my story.

There are three ways in which a tourist may obtain a good idea of Britain during a summer's vacation of three or four months. He may cover most places of interest after the old manner, by railway train. This will have to be supplemented by many and expensive carriage drives if he wishes to see the most beautiful country and many of the most interesting places. As Professor Goldwin Smith says, "Railways in England do not follow the lines of beauty in very many cases," and the opportunity afforded of really seeing England from a railway car window is poor indeed. The tourist must keep a constant eye on the time-tables, and in many of the more retired places he will have to spend a day when an hour would suffice quite as well could he get away. If he travels first-class, it is quite expensive, and the only advantage secured is that he generally has a compartment to himself, the difference in accommodations between first and third-class on the longer distance trains being insignificant. But if he travels third-class, he very often finds himself crowded into a small compartment with people in whom, to say the least, he has nothing in common. One seldom gets the real sentiment and beauty of a place in approaching it by railway. I am speaking, of course, of the tourist who endeavors to crowd as much as he can into a comparatively short time. To the one who remains several days in a place, railroad traveling is less objectionable. My remarks concerning railroad travel in England are made merely from the point of comparison with a pleasure journey by motor, and having covered the greater part of the country in both ways, I am qualified to some extent to speak from experience.

For a young man or party of young men who are traveling through Britain on a summer's vacation, the bicycle affords an excellent and expeditious method of getting over the country, and offers nearly all the advantages of the motor car, provided the rider is vigorous and expert enough to do the wheeling without fatigue. The motor cycle is still better from this point of view, and many thousands of them are in use on English roads, while cyclists may be counted by the tens of thousands. But the bicycle is out of the question for an extended tour by a party which includes ladies. The amount of impedimenta which must be carried along, and the many long hills which are encountered on the English roads, will put the cycle out of the question in such cases.

In the motor car, we have the most modern and thorough means of traversing the highways and byways of Britain in the limits of a single summer, and it is my purpose in this book, with little pretensions to literary style, to show how satisfactorily this may be done by a mere layman. To the man who drives his own car and who at the outstart knows very little about the English roads and towns, I wish to undertake to show how in a trip of five thousand miles, occupying about fifty days, actual traveling time, I covered much of the most beautiful country in England and Scotland and visited a large proportion of the most interesting and historic places in the Kingdom. I think it can be clearly demonstrated that this method of touring will give opportunities for enjoyment and for gaining actual knowledge of the people and country that can hardly be attained in any other way.

The motor car affords expeditious and reasonably sure means of getting over the country—always ready when you are ready, subservient to your whim to visit some inaccessible old ruin, flying over the broad main highways or winding more cautiously in the unfrequented country byways—and is, withal, a method of locomotion to which the English people have become tolerant if not positively friendly. Further, I am sure it will be welcome news to many that the expense of such a trip, under ordinary conditions, is not at all exorbitant or out of the reach of the average well-to-do citizen.

Those who have traveled for long distances on American roads can have no conception whatever of the delights of motor traveling on the British highways. I think there are more bad roads in the average county, taking the States throughout, than there are in all of the United Kingdom, and the number of defective bridges in any county outside of the immediate precincts of a few cities, would undoubtedly be many times greater than in the whole of Great Britain. I am speaking, of course, of the more traveled highways and country byways. There are roads leading into the hilly sections that would not be practicable for motors at all, but, fortunately, these are the very roads over which no one would care to go. While the gradients are generally easier than in the States, there are in many places sharp hills where the car must be kept well under control. But the beauty of it is that in Britain one has the means of being thoroughly warned in advance of the road conditions which he must encounter.

The maps are perfect to the smallest detail and drawn to a large scale, showing the relative importance of all the roads; and upon them are plainly marked the hills that are styled "dangerous." These maps were prepared for cyclists, and many of the hills seem insignificant to a powerful motor. However, the warning is none the less valuable, for often other conditions requiring caution prevail, such as a dangerous turn on a hill or a sharp descent into a village street. Then there is a set of books, four in number, published by an Edinburgh house and illustrated by profile plans, covering about thirty thousand miles of road in England and Scotland. These show the exact gradients and supply information in regard to the surface of the roads and their general characteristics. Besides this, the "objects of interest" scattered along any particular piece of road are given in brief—information at once so desirable and complete as to be a revelation to an American. There are sign-boards at nearly every crossing; only in some of the more retired districts did we find the crossroads unmarked. With such advantages as these, it is easily seen that a tour of Britain by a comparative stranger is not difficult; that a chauffeur or a guide posted on the roads is not at all necessary. The average tourist, with the exercise of ordinary intelligence and a little patience, can get about any part of the country without difficulty. One of the greatest troubles we found was to strike the right road in leaving a town of considerable size, but this was overcome by the extreme willingness of any policeman or native to give complete information—often so much in detail as to be rather embarrassing. The hundreds of people from whom we sought assistance in regard to the roads were without exception most cheerful and willing compliants, and in many places people who appeared to be substantial citizens volunteered information when they saw us stop at the town crossing to consult our maps. In getting about the country, little difficulty or confusion will be experienced.

Generally speaking, the hotel accommodations in the provincial towns throughout England and Scotland are surprisingly good. Of course there is a spice of adventure in stopping occasionally at one of the small wayside inns or at one of the old hostelries more famous for its associations than for comfort, but to one who demands first-class service and accommodations, a little of this will go a long way. Generally it can be so planned that towns with strictly good hotel accommodations can be reached for the night. Occasionally an unusually comfortable and well-ordered hotel will tempt the motorist to tarry a day or two and possibly to make excursions in the vicinity. Such hotels we found at Chester and York, for instance. The country hotel-keeper in Britain is waking up to the importance of motor travel. Already most of the hotels were prepared to take care of this class of tourists, and in many others improvements were under way. It is safe to say that in the course of two or three years, at the farthest, there will be little to be desired in the direction of good accommodations in the better towns. Rates at these hotels are not low by any means—at least for the motorist. It is generally assumed that a man who is in possession of an automobile is able to pay his bills, and charges and fees are exacted in accordance with this idea. There is, of course, a wide variation in this particular, and taking it right through, the rates at the best hotels would not be called exorbitant. The Motor Club of Great Britain and Ireland have many especially designated hotels where the members of this association are given a discount. These are not in every case the best in the town, and we generally found Baedeker's Hand Book the most reliable guide as to the relative merits of the hotels. It is a poorly appointed hotel that does not now have a garage of some sort, and in many cases, necessary supplies are available. Some even go so far as to charge the storage batteries, or "accumulators," as they are always called in Britain, and to afford facilities for the motorist to make repairs.

It goes without saying that a motor tour should be planned in advance as carefully as possible. If one starts out in a haphazard way, it takes him a long time to find his bearings, and much valuable time is lost. Before crossing the water, it would be well to become posted as thoroughly as possible on what one desires to see and to gain a general idea of the road from the maps. Another valuable adjunct will be a membership in the A.C.A. or a letter from the American motor associations, with an introduction to the Secretary of the Motor Union of Great Britain and Ireland. In this manner can be secured much valuable information as to the main traveled routes; but after all, if the tourist is going to get the most out of his trip, he will have to come down to a careful study of the country and depend partly on the guide-books but more upon his own knowledge of the historical and literary landmarks throughout the Kingdom.

II

IN AND ABOUT LONDON

London occurs to the average tourist as the center from which his travels in the Kingdom will radiate, and this idea, from many points of view, is logically correct. Around the city cluster innumerable literary and historic associations, and the points of special interest lying within easy reach will outnumber those in any section of similar extent in the entire country. If one purposes to make the tour by rail, London is pre-eminently the center from which to start and to which one will return at various times in his travels. All the principal railways lead to the metropolis. The number of trains arriving and departing each day greatly exceeds that of any other city in the world, and the longest through journey in the island may be compassed between sunrise and sunset.

The motorist, however, finds a different problem confronting him in making London his center. I had in mind the plan of visiting the famous places of the city and immediate suburbs with the aid of my car, but it was speedily abandoned when I found myself confronted by the actual conditions. One attempt at carrying out this plan settled the matter for me. The trip which I undertook would probably be one of the first to occur to almost anybody—the drive to Hampton Court Palace, about twelve or fifteen miles from the central part of the city. It looked easy to start about two or three o'clock, spend a couple of hours at Hampton Court and get back to our hotel by six. After trying out my car—which had reached London some time ahead of me—a few times in localities where traffic was not the heaviest, I essayed the trip without any further knowledge of the streets than I had gained from the maps. I was accompanied by a nervous friend from Iowa who confessed that he had been in an automobile but once before. He had ridden with a relative through a retired section of his native state, traversed for the first time by an automobile, and he had quit trying to remember how many run-aways and smash-ups were caused by the fractious horses they met on the short journey. Visions of damage suits haunted him for months thereafter. In our meanderings through the London streets, the fears for the other fellow which had harassed him during his former experience, were speedily transferred to himself. To his excited imagination, we time and again escaped complete wreck and annihilation by a mere hair's breadth. The route which we had taken, I learned afterwards, was one of the worst for motoring in all London. The streets were narrow and crooked and were packed with traffic of all kinds. Tram cars often ran along the middle of the street, with barely room for a vehicle to pass on either side. The huge motor busses came tearing towards us in a manner most trying to novices, and it seemed, time after time, that the dexterity of the drivers of these big machines was all that saved our car from being wrecked. We obtained only the merest glimpse of Hampton Palace, and the time which we had consumed made it apparent that if we expected to reach our hotel that night, we must immediately retrace our way through the wild confusion we had just passed. It began to rain, and added to the numerous other dangers that seemed to confront us was that of "skidding" on the slippery streets. When we finally reached our garage, I found that in covering less than twenty-five miles, we had consumed about four hours and we had been moving all the time. The nervous strain was a severe one and I forthwith abandoned any plan that I had of attempting to do London by motor car. With more knowledge and experience I would have done better, but a local motorist, thoroughly acquainted with London, told me that he wouldn't care to undertake the Hampton Court trip by the route which we had traveled.

On Saturday afternoons and Sundays, the motorist may practically have freedom of the city. He will find the streets deserted everywhere. The heavy traffic has all ceased and the number of cabs and motor busses is only a fraction of what it would be on business days. He will meet comparatively few motors in the city on Sunday, even though the day be fine, such as would throng the streets of Chicago or New York with cars. The Englishman who goes for a drive is attracted from the city by the many fine roads which lead in every direction to pleasure resorts. One of the most popular runs with Londoners is the fifty miles to Brighton, directly southward, and the number of motors passing over this highway on fine Sundays is astonishing. I noted a report in the papers that on a certain Sunday afternoon no less than two hundred cars passed a police trap, and of these, thirty-five were summoned before the magistrates for breaking the speed limit. To the average American, this run to Brighton would not be at all attractive compared with many other roads leading out of London, on which one would scarcely meet a motor car during the day and would be in no danger from the machinations of the police. Of course the places frequented by tourists are often closed on Sunday—or at least partially so, as in the case of Windsor Castle, where one is admitted to the grounds and court, but the state apartments, etc., are not shown. Even the churches are closed to Sunday visitors except during the regular services.

Within a radius of thirty miles of London, and outside its immediate boundaries, there are numerous places well worth a visit, most of them open either daily or at stated times. A few of such places are Harrow on the Hill, with its famous school; Keston, with Holwood House, the home of William Pitt; Chigwell, the scene of Dickens' "Barnaby Rudge;" Waltham Abbey Church, founded in 1060; the home of Charles Darwin at Downe; Epping Forest; Hampton Court; Rye House at Broxborne; Hatfield House, the estate of the Marquis of Salisbury; Runnymede, where the Magna Charta was signed; St. Albans, with its ancient cathedral church; Stoke Poges Church of Gray's "Elegy" fame; Windsor Castle; Knole House, with its magnificent galleries and furniture; Penshurst Place, the home of the Sidneys; John Milton's cottage at Chalfont St. Giles; the ancient town of Guildford in Surrey; Gad's Hill, Dickens' home, near Rochester; the vicarage where Thackeray's grandfather lived and the old church where he preached at Monken Hadley; and Whitchurch, with Handel's original organ, is also near the last-named village. These are only a few of the places that no one should miss. The motor car affords an unequalled means of reaching these and other points in this vicinity; since many are at some distance from railway stations, to go by train would consume more time than the average tourist has at his disposal. While we visited all the places which I have just mentioned and many others close to London, we made only three or four short trips out of the city returning the same or the following day. We managed to reach the majority of such points by going and returning over different highways on our longer tours. In this way we avoided the difficulty we should have experienced in making many daily trips from London, since a large part of each day would have been consumed merely in getting in and out of the city.

Our first trip into the country was made on the Sunday after our arrival. Although we started out at random, our route proved a fortunate one, and gave us every reason to believe that our tour of the Kingdom would be all we had anticipated. During the summer we had occasion to travel three times over this same route, and we are still of the opinion that there are few more delightful bits of road in England. We left London by the main highway, running for several miles through Epping Forest, which is really a great suburban park. It was a good day for cyclists, for the main road to the town of Epping was crowded with thousands of them. So great was the number and so completely did they occupy the highway, that it was necessary to drive slowly and with the greatest care. Even then, we narrowly avoided a serious accident. One of the cyclists, evidently to show his dexterity, undertook to cut around us by running across the tramway tracks. These were wet and slippery, and the wheel shot from under the rider, pitching him headlong to the ground not two feet in front of our car, which was then going at a pretty good rate. If the cyclist did not exhibit skill in managing his wheel, he certainly gave a wonderful display of agility in getting out of our way. He did not seem to touch the ground at all, and by turning two or three handsprings, he avoided being run over by the narrowest margin. His wheel was considerably damaged and his impedimenta scattered over the road. It was with rather a crestfallen air that he gathered up his belongings, and we went on, shuddering to think how close we had come to a serious accident at the very beginning of our pilgrimage. A policeman witnessed the accident, but he clearly placed the blame on the careless wheelman.

Passing through the forest, we came to Epping, and from there into a stretch of open country that gave little suggestion of proximity to the world's metropolis. Several miles through a narrow but beautifully kept byway brought us to the village of Chipping-Ongar, a place of considerable antiquity, and judging from the extensive site of its ancient castle, at one time of some military importance.

At Ongar we began our return trip to London over the road which we agreed was the most beautiful leading out of the city, for the suburbs do not extend far in this direction and one is comparatively soon in the country. The perfectly surfaced road, with only gentle slopes and curves, runs through the parklike fields, here over a picturesque stone bridge spanning a clear stream, there between rows of magnificent trees, occasionally dropping into quiet villages, of which Chigwell was easily the most delightful.

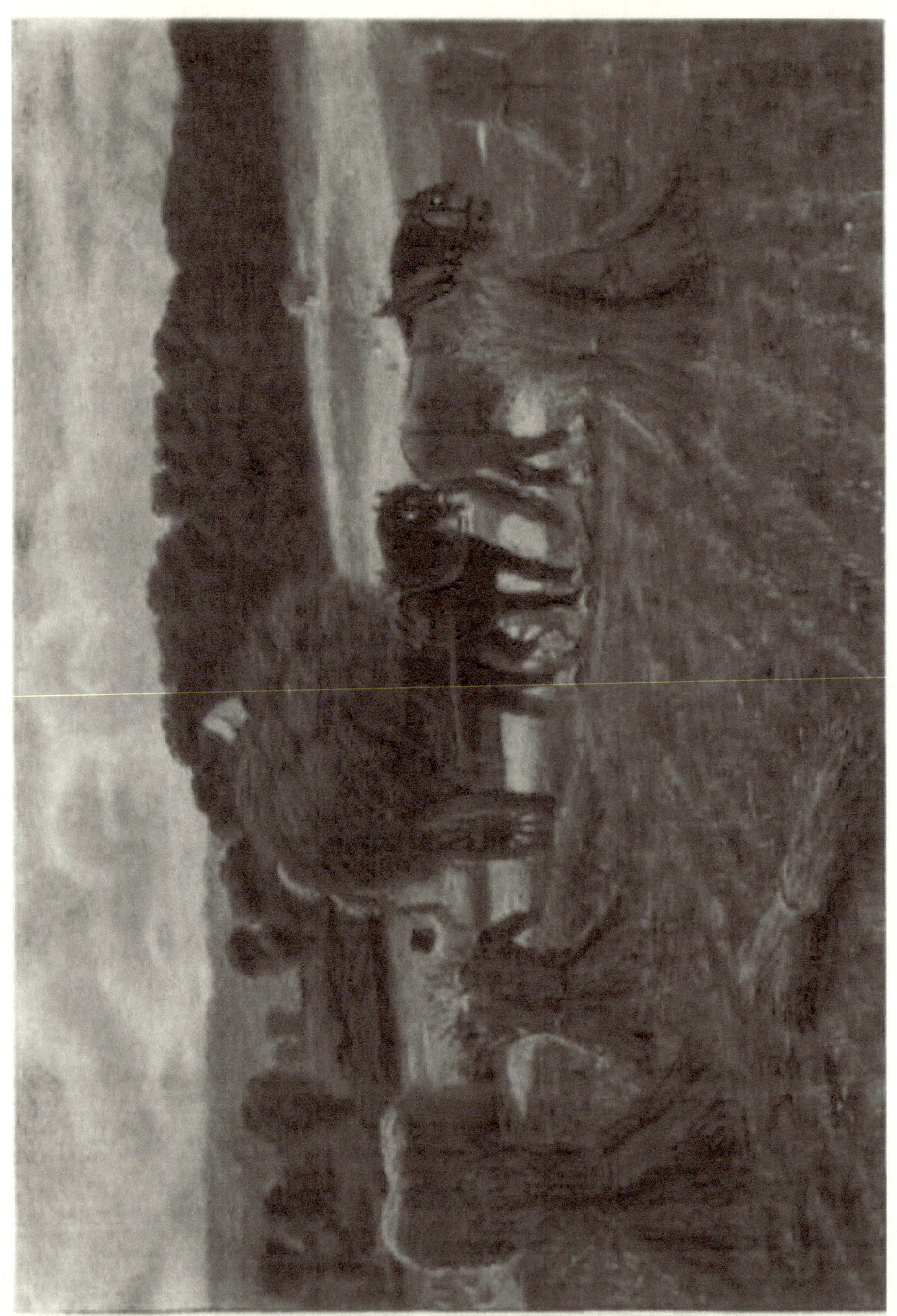

HARVESTING IN HERTFORDSHIRE.

From Painting by Alfred Elias. Exhibited in 1906 Royal Academy.

Chigwell became known to fame through the writings of Charles Dickens, who was greatly enamored of the place and who made it the scene of much of his story of "Barnaby Rudge." But Dickens, with his eye for the beautiful and with his marvelous intuition for interesting situations, was drawn to the village by its unusual charm. Few other places can boast of such endorsement as he gave in a letter to his friend, Forster, when he wrote: "Chigwell, my dear fellow, is the greatest place in the world. Name your day for going. Such a delicious old inn facing the church; such a lovely ride; such glorious scenery; such an out-of-the-way rural place; such a sexton! I say again, name your day." After such a recommendation, one will surely desire to visit the place, and it is pleasant to know that the "delicious old inn" is still standing and that the village is as rural and pretty as when Dickens wrote over sixty years since.

The inn referred to, the King's Head, was the prototype of the Maypole in "Barnaby Rudge," and here we were delighted to stop for our belated luncheon. The inn fronts directly on the street and, like all English hostelries, its main rooms are given over to the bar, which at this time was crowded with Sunday loafers, the atmosphere reeking with tobacco smoke and the odor of liquors. The garden at the rear was bright with a profusion of spring flowers and sheltered with ornamental trees and vines. The garden side of the old house was covered with a mantle of ivy, and, altogether, the surroundings were such as to make ample amends for the rather unprepossessing conditions within. One will not fully appreciate Chigwell and its inn unless he has read Dickens' story. You may still see the panelled room upstairs where Mr. Chester met Geoffry Haredale. This room has a splendid mantel-piece, great carved open beams and beautiful leaded windows. The bar-room, no doubt, is still much the same as on the stormy night which Dickens chose for the opening of his story. Just across the road from the inn is the church which also figures in the tale, and a dark avenue of ancient yew trees leads from the gateway to the door. One can easily imagine the situation which Dickens describes when the old sexton crossed the street and rang the church bells on the night of the murder at Haredale Hall.

Aside from Dickens' connection with Chigwell, the village has a place of peculiar interest to Americans in the old grammar school where William Penn received his early education. The building still stands, with but little alteration, much as it was in the day when the great Quaker sat at the rude desks and conned the lessons of the old-time English schoolboy.

When we invited friends whom we met in London to accompany us on a Sunday afternoon trip, we could think of no road more likely to please them than the one I have just been trying to describe. We reversed our journey this time, going out of London on the way to Chigwell. Returning, we left the Epping road shortly after passing through that town, and followed a narrow, forest-bordered byway with a few steep hills until we came to Waltham Abbey, a small Essex market town with an important history. The stately abbey church, a portion of which is still standing and now used for services, was founded by the Saxon king, Harold, in 1060. Six years later he was defeated and slain at Hastings by William the Conqueror, and tradition has it that his mother buried his body a short distance to the east of Waltham Church. The abbey gate still stands as a massive archway at one end of the river bridge. Near the town is one of the many crosses erected by Edward I in memory of his wife, Eleanor of Castile, wherever her body rested on the way from Lincoln to Westminster. A little to the left of this cross, now a gateway to Theobald Park, stands Temple Bar, stone for stone intact as it was in the days when traitors' heads were raised above it in Fleet Street, although the original wooden gates are missing. Waltham Abbey is situated on the River Lea, near the point where King Alfred defeated the Danes in one of his battles. They had penetrated far up the river when King Alfred diverted the waters from beneath their vessels and left them stranded in a wilderness of marsh and forest.

Another pleasant afternoon trip was to Monken Hadley, twenty-five miles out on the Great North Road. Hadley Church is intimately associated with a number of distinguished literary men, among them Thackeray, whose grandfather preached there and is buried in the churchyard. The sexton was soon found and he was delighted to point out the interesting objects in the church and vicinity.

The church stands at the entrance of a royal park, which is leased to private parties and is one of the quaintest and most picturesque of the country churches we had seen. Over the doors, some old-fashioned figures which we had to have translated indicated that the building had been erected in 1494. It has a huge ivy-covered tower and its interior gives every evidence of the age-lasting solidity of the English churches.

Hadley Church has a duplicate in the United States, one having been built in some New York town precisely like the older structure. We noticed that one of the stained-glass windows had been replaced by a modern one, and were informed that the original had been presented to the newer church in America—a courtesy that an American congregation would hardly think of, and be still less likely to carry out. An odd silver communion service which had been in use from three to five hundred years was carefully taken out of a fire-proof safe and shown us.

Hadley Church is a delight from every point of view, and it is a pity that such lines of architecture are not oftener followed in America. Our churches as a rule are shoddy and inharmonious affairs compared with those in England. It is not always the matter of cost that makes them so, since more artistic structures along the pleasing and substantial lines of architecture followed in Britain would in many cases cost no more than we pay for such churches as we now have.

Our friend the sexton garrulously assured us that Thackeray had spent much of his time as a youth at the vicarage and insisted that a great part of "Vanity Fair" was written there. He even pointed out the room in which he alleged the famous book was produced, and assured us that the great author had found the originals of many of his characters, such as Becky Sharp and Col. Newcome, among the villagers of Hadley. All of which we took for what it was worth. Thackeray himself told his friend, Jas. T. Fields, that "Vanity Fair" was written in his London house; still, he may have been a visitor at the Hadley vicarage and might have found pleasure in writing in the snug little room whose windows open on the flower garden, rich with dashes of color that contrasted effectively with the dark green foliage of the hedges and trees. The house still does duty as a vicarage; the small casement windows peep out of the ivy that nearly envelops it, and an air of coziness and quiet seems to surround it. Near at hand is the home where Anthony Trollope, the novelist, lived for many years, and his sister is buried in the churchyard.

HADLEY CHURCH, MONKEN HADLEY.

A short distance from Hadley is the village of Edgeware, with Whitchurch, famous for its association with the musician Handel. He was organist here for several years, and on the small pipe-organ, still in the church though not in use, composed his oratorio, "Esther," and a less important work, "The Harmonious Blacksmith." The idea of the latter came from an odd character, the village blacksmith, who lived in Edgeware in Handel's day and who acquired some fame as a musician. His tombstone in the churchyard consists of an anvil and hammer, wrought in stone. Afterwards Handel became more widely known, and was called from Whitchurch for larger fields of work. He is buried in Westminster Abbey.

The road from Edgeware to the city is a good one, and being Saturday afternoon, it was nearly deserted. Saturday in London is quite as much of a holiday as Sunday, little business being transacted, especially in the afternoon. This custom prevails to a large extent all over the Kingdom, and rarely is any attempt made to do business on Saturday. The Week-End holiday, as it is called, is greatly prized, and is recognized by the railroads in granting excursions at greatly reduced rates. There is always a heavy exodus of people from the city to the surrounding resorts during the summer and autumn months on Saturday afternoon and Sunday.

Owing to the extreme difficulty of getting about the city, we made but few short excursions from London such as I have described. If one desires to visit such places in sequence, without going farther into the country, it would be best to stop for the night at the hotels in the better suburban towns, without attempting to return to London each day.

The garage accomodations in London I found very good and the charges generally lower than in the United States. There is a decided tendency at grafting on the part of the employes, and if it is ascertained that a patron is a tourist—especially an American—he is quoted a higher rate at some establishments and various exactions are attempted. At the first garage where I applied, a quotation made was withdrawn when it was learned that I was an American. The man said he would have to discuss the matter with his partner before making a final rate. I let him carry on his discussion indefinitely, for I went on my way and found another place where I secured accommodations at a very reasonable rate without giving information of any kind.

With the miserable business methods in vogue at some of the garages, it seemed strange to me if any of the money paid to employes ever went to the business office at all. There was no system and little check on sales of supplies, and I heard a foreman of a large establishment declare that he had lost two guineas which a patron had paid him. "I can't afford to lose it," he said, "and it will have to come back indirectly if I can't get it directly." In no case should a motorist pay a bill at a London garage without a proper receipt.

III

A PILGRIMAGE TO CANTERBURY

No place within equal distance of London is of greater interest than Canterbury, and, indeed, there are very few cities in the entire Kingdom that can vie with the ancient cathedral town in historical importance and antiquity. It lies only sixty-five miles southeast of London, but allowing for the late start that one always makes from an English hotel, and the points that will engage attention between the two cities, the day will be occupied by the trip. Especially will this be true if, as in our case, fully two hours be spent in getting out of the city and reaching the highway south of the Thames, which follows the river to Canterbury.

Leaving Russell Square about ten o'clock, I followed the jam down Holborn past the Bank and across London Bridge, crawling along at a snail's pace until we were well beyond the river. A worse route and a more trying one it would have been hard to select. With more experience, I should have run down the broad and little-congested Kingsway to Waterloo Bridge and directly on to Old Kent road in at least one-fourth the time which I consumed in my ignorance. Nevertheless, if a novice drives a car in London, he can hardly avoid such experiences. Detailed directions given in advance cannot be remembered and there is little opportunity to consult street signs and maps or even to question the policeman in the never-ending crush of the streets. However, one gradually gains familiarity with the streets and landmarks, and by the time I was ready to leave London for America, I had just learned to get about the city with comparative ease.

Old Kent road, which leads out of London towards Canterbury, is an ancient highway, and follows nearly, if not quite, the route pursued by the Canterbury pilgrims of the poet Chaucer. In the main it is unusually broad and well kept, but progress will be slow at first, as the suburbs extend a long way in this direction, and for the first twenty-five miles one can hardly be said to be out of the city at any time. Ten miles out the road passes Greenwich, where the British observatory is located, and Woolwich, the seat of the great government arsenals and gun works, is also near this point, lying directly by the river.

Nearly midway between London and Rochester is the old town of Dartford, where we enjoyed the hospitality of the Bull Hotel for luncheon. A dingy, time-worn, rambling old hostelry it is, every odd corner filled with stuffed birds and beasts to an extent that suggested a museum, and as if to still further carry out the museum feature, mine host had built in a small court near the entrance a large cage or bird-house which was literally alive with specimens of feathered songsters of all degrees. The space on the first floor not occupied by these curios was largely devoted to liquor selling, for there appeared to be at least three bars in the most accessible parts of the hotel. However, somewhat to the rear there was a comfortable coffee room, where our luncheon was neatly served. We had learned by this time that all well regulated hotels in the medium sized towns, and even in some of the larger cities—as large as Bristol, for instance—have two dining rooms, one, generally for tourists, called the "coffee room," with separate small tables, and a much larger room for "commercials," or traveling salesmen, where all are seated together at a single table. The service is practically the same, but the ratio of charges is from two to three times higher in the coffee room. We found many old hotels in retired places where a coffee room had been hastily improvised, an innovation no doubt brought about largely by the motor car trade and the desire to give the motorist more aristocratic rates than those charged the well-posted commercials. Though we stopped in Dartford no longer than necessary for lunch and a slight repair to the car, it is a place of considerable interest. Its chief industry is a large paper-mill, a direct successor to the first one established in England near the end of the Sixteenth Century, and Foolscap paper, standard throughout the English-speaking world, takes its name from the crest (a fool's cap) of the founder of the industry, whose tomb may still be seen in Dartford Church.

A short run over a broad road bordered with beautiful rural scenery brought us into Rochester, whose cathedral spire and castle with its huge Norman tower loomed into view long before we came into the town itself. A few miles out of the town our attention had been attracted by a place of unusual beauty, a fine old house almost hidden by high hedges and trees on one side of the road and just opposite a tangled bit of wood and shrubbery, with several of the largest cedars we saw in England. So picturesque was the spot that we stopped for a photograph of the car and party, with the splendid trees for a background, but, as often happens in critical cases, the kodak film only yielded a "fog" when finally developed.

When we reached Rochester, a glance at the map showed us that we had unwittingly passed Gad's Hill, the home where Charles Dickens spent the last fifteen years of his life and where he died thirty-six years ago. We speedily retraced the last four or five miles of our journey and found ourselves again at the fine old place with the cedar trees where we had been but a short time before. We stopped to inquire at a roadside inn which, among the multitude of such places, we had hardly noticed before, and which bore the legend, "The Sir John Falstaff," a distinction earned by being the identical place where Shakespeare located some of the pranks of his ridiculous hero. The inn-keeper was well posted on the literary traditions of the locality. "Yes," said he, "this is Gad's Hill Place, where Dickens lived and where he died just thirty-six years ago today, on June 9th, 1870; but the house is shown only on Wednesdays of each week and the proprietor doesn't fancy being troubled on other days. But perhaps, since you are Americans and have come a long way, he may admit you on this special anniversary. Anyway, it will do no harm for you to try."

DICKENS' HOME, GAD'S HILL, NEAR ROCHESTER.

Personally, I could not blame the proprietor for his disinclination to admit visitors on other than the regular days, and it was impressed on me more than once during our trip that living in the home of some famous man carries quite a penalty, especially if the present owner happens to be a considerate gentleman who dislikes to deprive visitors of a glimpse of the place. Such owners are often wealthy and the small fees which they fix for admittance are only required as evidence of good faith and usually devoted to charity. With a full appreciation of the situation, it was not always easy to ask for the suspension of a plainly stated rule, yet we did this in many instances before our tour was over and almost invariably with success. In the present case we were fortunate, for the gentleman who owned Gad's Hill was away and the neat maid who responded to the bell at the gateway seemed glad to show us the place, regardless of rules. It is a comfortable, old-fashioned house, built about 1775, and was much admired by Dickens as a boy when he lived with his parents in Rochester. His father used to bring him to look at the house and told him that if he grew up a clever man, he might possibly own it some time.

We were first shown into the library, which is much the same as the great writer left it at his death, and the chair and desk which he used still stand in their accustomed places. The most curious feature of the library is the rows of dummy books that occupy some of the shelves, and even the doors are lined with these sham leather backs glued to boards, a whim of Dickens carefully respected by the present owner. We were also accorded a view of the large dining room where Dickens was seized with the attack which resulted in his sudden and unexpected death. After a glimpse of other parts of the house and garden surrounding it, the maid conducted us through an underground passage leading beneath the road, to the plot of shrubbery which lay opposite the mansion. In this secluded thicket, Dickens had built a little house, to which in the summer time he was often accustomed to retire when writing. It was an ideal English June day, and everything about the place showed to the best possible advantage. We all agreed that Gad's Hill alone would be well worth a trip from London. The country around is surpassingly beautiful and it is said that Dickens liked nothing better than to show his friends about the vicinity. He thought the seven miles between Rochester and Maidstone the most charming walk in all England. He delighted in taking trips with his friends to the castles and cathedrals and he immensely enjoyed picnics and luncheons in the cherry orchards and gardens.

A very interesting old city is Rochester, with its Eleventh Century cathedral and massive castle standing on the banks of the river. Little of the latter remains save the square tower of the Norman keep, one of the largest and most imposing we saw in England. The interior had been totally destroyed by fire hundreds of years ago, but the towering walls of enormous thickness still stand firm. Its antiquity is attested by the fact that it sustained a siege by William Rufus, the son of the Conqueror. The cathedral is not one of the most impressive of the great churches. It was largely rebuilt in the Twelfth Century, the money being obtained from miracles wrought by the relics of St. William of Perth, a pilgrim who was murdered on his way to Canterbury and who lies buried in the cathedral. Rochester is the scene of many incidents of Dickens' stories. It was the scene of his last unfinished work, "Edwin Drood," and he made many allusions to it elsewhere, the most notable perhaps in "Pickwick Papers," where he makes the effervescent Mr. Jingle describe it thus: "Ah, fine place, glorious pile, frowning walls, tottering arches, dark nooks, crumbling staircases—old cathedral, too,—earthy smell—pilgrims' feet worn away the old steps."

Across the river from Rochester lies Chatham, a city of forty thousand people and a famous naval and military station. The two cities are continuous and practically one. From here, without further stop, we followed the fine highway to Canterbury and entered the town by the west gate of Chaucer's Tales. This alone remains of the six gateways of the city wall in the poet's day, and the strong wall itself, with its twenty-one towers, has almost entirely disappeared. We followed a winding street bordered with quaint old buildings until we reached our hotel—in this case a modern and splendidly kept hostelry. The hotel was just completing an extensive garage, but it was not ready for occupancy and I was directed to a well equipped private establishment with every facility for the care and repair of motors. The excellence of the service at this hotel attracted our attention and the head waiter told us that the owners had their own farm and supplied their own table—accounting in this way for the excellence and freshness of the milk, meat and vegetables.

The long English summer evening still afforded time to look about the town after dinner. Passing down the main street after leaving the hotel, we found that the river and a canal wound their way in several places between the old buildings closely bordering on each side. The whole effect was delightful and so soft with sunset colors as to be suggestive of Venice. We noted that although Canterbury is exceedingly ancient, it is also a city of nearly thirty thousand population and the center of rich farming country, and, as at Chester, we found many evidences of prosperity and modern enterprise freely interspersed with the quaint and time-worn landmarks. One thing which we noticed not only here but elsewhere in England was the consummate architectural taste with which the modern business buildings were fitted in with the antique surroundings, harmonizing in style and color, and avoiding the discordant note that would come from a rectangular business block such as an American would have erected. Towns which have become known to fame and to the dollar-distributing tourists are now very slow to destroy or impair the old monuments and buildings that form their chief attractiveness, and the indifference that prevailed generally fifty or a hundred years ago has entirely vanished. We in America think we can afford to be iconoclastic, for our history is so recent and we have so little that commands reverence by age and association; yet five hundred years hence our successors will no doubt bitterly regret this spirit of their ancestors, just as many ancient towns in Britain lament the folly of their forbears who converted the historic abbeys and castles into hovels and stone fences.

Fortunately, the cathedral at Canterbury escaped such a fate, and as we viewed it in the fading light we received an impression of its grandeur and beauty that still keeps it pre-eminent after having visited every cathedral in the island. It is indeed worthy of its proud position in the English church and its unbroken line of traditions, lost in the mist of antiquity. It is rightly the delight of the architect and the artist, but an adequate description of its magnificence has no place in this hurried record. Time has dealt gently with it and careful repair and restoration have arrested its decay. It stands today, though subdued and stained by time, as proudly as it did when a monarch, bare-footed, walked through the roughly paved streets to do penance at the tomb of its martyred archbishop. It escaped lightly during the Reformation and civil war, though Becket's shrine was despoiled as savoring of idolatry and Cromwell's men desecrated its sanctity by stabling their horses in the great church.

The next day being Sunday, we were privileged to attend services at the cathedral, an opportunity we were always glad to have at any of the cathedrals despite the monotony of the Church of England service, for the music of the superb organs, the mellowed light from the stained windows, and the associations of the place were far more to us than litany or sermon. The archbishop was present at the service in state that fitted his exalted place as Primate of all England and his rank, which, as actual head of the church, is next to the king, nominally head of the church as well as of the state. He did not preach the sermon but officiated in the ordination of several priests, a service full of solemn and picturesque interest. The archbishop was attired in his crimson robe of state, the long train of which was carried by young boys in white robes, and he proceeded to his throne with all the pomp and ceremony that so delights the soul of the Englishman. He was preceded by several black-robed officials bearing the insignia of their offices, and when he took his throne, he became apparently closely absorbed in the sermon, which was preached by a Cambridge professor.

We were later astonished to learn that the archbishop's salary amounts to $75,000 per year, or half as much more than that of the President of the United States, and we were still more surprised to hear that the heavy demands made on him in maintaining his state and keeping up his splendid episcopal palaces are such that his income will not meet them. We were told that the same situation prevails everywhere with these high church dignitaries, and that only recently the Bishop of London had published figures to show that he was $25,000 poorer in the three years of his incumbency on an annual salary of $40,000 per year. It is not strange, therefore, that among these churchmen there exists a demand for a simpler life. The Bishop of Norwich frankly acknowledged recently that he had never been able to live on his income of $22,500 per year. He expressed his conviction that the wide-spread poverty of the bishops is caused by their being required to maintain "venerable but costly palaces." He says that he and many of his fellow-churchmen would prefer to lead plain and unostentatious lives, but they are not allowed to do so; that they would much prefer to devote a portion of their income to charity and other worthy purposes rather than to be compelled to spend it in useless pomp and ceremony.

CATHEDRAL, CANTERBURY.

Aside from its cathedral, Canterbury teems with unique relics of the past, some antedating the Roman invasion of England. The place of the town in history is an important one, and Dean Stanley in his "Memorials of Canterbury," claims that three great landings were made in Kent adjacent to the city, "that of Hengist and Horsa, which gave us our English forefathers and character; that of Julius Caesar, which revealed to us the civilized world, and that of St. Augustine, which gave us our Latin Christianity." The tower of the cathedral dominates the whole city and the great church often overshadows everything else in interest to the visitor. But one could spend days in the old-world streets, continually coming across fine half-timbered houses, with weather-beaten gables in subdued colors and rich antique oak carvings. There are few more pleasing bits of masonry in Britain than the great cathedral gateway at the foot of Mercery Lane, with its rich carving, weather-worn to a soft blur of gray and brown tones. Near Mercery Lane, too, are slight remains of the inn of Chaucer's Tales, "The Chequers of Hope," and in Monastery Street stands the fine gateway of the once rich and powerful St. Augustine's Abbey. Then there is the quaint little church of St. Martins, undoubtedly one of the oldest in England, and generally reputed to be the oldest. Here, in the year 600, St. Augustine preached before the cathedral was built. Neither should St. John's hospital, with its fine, half-timbered gateway be forgotten; nor the old grammar school, founded in the Seventh Century.

Our stay in the old town was all too short, but business reasons demanded our presence in London on Monday, so we left for that city about two o'clock. We varied matters somewhat by taking a different return route, and we fully agreed that the road leading from Canterbury to London by way of Maidstone is one of the most delightful which we traversed in England. It led through fields fresh with June verdure, losing itself at times in great forests, where the branches of the trees formed an archway overhead. Near Maidstone we caught a glimpse of Leeds Castle, one of the finest country seats in Kent, the main portions of the building dating from the Thirteenth Century. We had a splendid view from the highway through an opening in the trees of the many-towered old house surrounded by a shimmering lake, and gazing on such a scene under the spell of an English June day, one might easily forget the present and fancy himself back in the time when knighthood was in flower, though the swirl of a motor rushing past us would have dispelled any such reverie had we been disposed to entertain it. We reached London early, and our party was agreed that our pilgrimage to Canterbury could not very well have been omitted from our itinerary.

IV

A RUN THROUGH THE MIDLANDS

I had provided myself with letters of introduction from the American Automobile Association and Motor League, addressed to the secretary of the Motor Union of Great Britain and Ireland, and shortly after my arrival in London, I called upon that official at the club headquarters. After learning my plans, he referred me to Mr. Maroney, the touring secretary, whom I found a courteous gentleman, posted on almost every foot of road in Britain and well prepared to advise one how to get the most out of a tour. Ascertaining the time I proposed to spend and the general objects I had in view, he brought out road-maps of England and Scotland and with a blue pencil rapidly traced a route covering about three thousand miles, which he suggested as affording the best opportunity of seeing, in the time and distance proposed, many of the most historic and picturesque parts of Britain.

In a general way, this route followed the coast from London to Land's End, through Wales north to Oban and Inverness, thence to Aberdeen and back to London along the eastern coast. He chose the best roads with unerring knowledge and generally avoided the larger cities. On the entire route which he outlined, we found only one really dangerous grade—in Wales—and, by keeping away from cities, much time and nervous energy were saved. While we very frequently diverged from this route, it was none the less of inestimable value to us, and other information, maps, road-books, etc., which were supplied us by Mr. Maroney, were equally indispensable. I learned that the touring department of the Union not only affords this service for Great Britain, but has equal facilities for planning tours in any part of Europe. In fact, it is able to take in hand the full details, such as providing for transportation of the car to some port across the Channel, arranging for necessary licenses and supplying maps and road information covering the different countries of Europe which the tourist may wish to visit. This makes it very easy for a member of the Union—or anyone to whom it may extend its courtesies—to go direct from Britain for a continental trip, leaving the tourist almost nothing to provide for except the difficulties he would naturally meet in the languages of the different countries.

When I showed a well posted English friend the route that had been planned, he pronounced very favorably upon it, but declared that by no means should we miss a run through the Midlands. He suggested that I join him in Manchester on business which we had in hand, allowing for an easy run of two days to that city by way of Coventry. On our return trip, we planned to visit many places not included in our main tour, among them the Welsh border towns, Shrewsbury and

Ludlow, and to run again through Warwickshire, taking in Stratford and Warwick, on our return to London. This plan was adopted and we left London about noon, with Coventry, nearly one hundred miles away, as our objective point.

A motor car is a queer and capricious creature. Before we were entirely out of the crush of the city, the engine began to limp and shortly came to a stop. I spent an hour hunting the trouble, to the entertainment and edification of the crowd of loafers who always congregate around a refractory car. I hardly know to this minute what ailed the thing, but it suddenly started off blithely, and this was the only exhibition of sulkiness it gave, for it scarcely missed a stroke in our Midland trip of eight hundred miles—mostly in the rain. Nevertheless, the little circumstance, just at the outset of our tour, was depressing.

We stopped for lunch at the Red Lion in the old town of St. Albans, twenty miles to the north of London. It is a place of much historic interest, being a direct descendant of the ancient Roman city of Verulamium; and Saint Albans, or Albanus, who gave his name to the town and cathedral and who was beheaded near this spot, was the first British martyr to Christianity of whom there is any record. The cathedral occupies the highest site of any in England, and the square Norman tower, which owes its red coloring to the Roman brick used in its construction, is a conspicuous object from the surrounding country. The nave is of remarkable length, being exceeded only by Winchester. Every style of architecture is represented, from early Norman to late Perpendicular, and there are even a few traces of Saxon work. The destruction of this cathedral was ordered by the pious Henry VIII at the time of his Reformation, but he considerately rescinded the order when the citizens of St. Albans raised money by public subscription to purchase the church. Only an hour was given to St. Albans, much less than we had planned, but our late start made it imperative that we move onward.

Our route for the day was over the old coach road leading from London to Holyhead, one of the most perfect in the Kingdom, having been in existence from the time of the Romans. In fact, no stretch of road of equal distance in our entire tour was superior to the one we followed from St. Albans to Coventry. It was nearly level, free from sharp turns, with perfect surface, and cared for with neatness such as we would find only in a millionaire's private grounds in the United States. Everywhere men were at work repairing any slight depression, trimming the lawnlike grasses on each side to an exact line with the edges of the stone surface, and even sweeping the road in many places to rid it of dust and dirt. Here and there it ran for a considerable distance through beautiful avenues of fine elms and yews; the hawthorne hedges which bordered it almost everywhere were trimmed with careful exactness; and yet amid all this precision there bloomed in many places the sweet English wild flowers—forget-me-nots, violets, wild hyacinths and bluebells. The country itself was rather flat and the villages generally uninteresting. The road was literally bordered with wayside inns, or, more properly, ale houses, for they apparently did little but sell liquor, and their names were odd and fantastic in a high degree. We noted a few of them. The "Stump and Pie," the "Hare and Hounds," the "Plume of Feathers," the "Blue Ball Inn," the "Horse and Wagon," the "Horse and Jockey," the "Dog and Parson," the "Dusty Miller," the "Angel Hotel" the "Dun Cow Inn," the "Green Man," the "Adam and Eve," and the "Coach and Horses," are a few actual examples of the fearful and wonderful nomenclature of the roadside houses. Hardly less numerous than these inns were the motor-supply depots along this road. There is probably no other road in England over which there is greater motor travel, and supplies of all kinds are to be had every mile or two. The careless motorist would not have far to walk should he neglect to keep up his supply of petrol—or motor spirit, as they call it everywhere in Britain.

Long before we reached Coventry, we saw the famous "three spires" outlined against a rather threatening cloud, and just as we entered the crooked streets of the old town, the rain began to fall heavily. The King's Head Hotel was comfortable and up-to-date, and the large room given us, with its fire burning brightly in the open grate, was acceptable indeed after the drive in the face of a sharp wind, which had chilled us through. And, by the way, there is little danger of being supplied with too many clothes and wraps when motoring in Britain. There were very few days during our entire summer's tour when one could dispense with cloaks and overcoats.

Coventry, with its odd buildings and narrow, crowded streets, reminded Nathaniel Hawthorne of Boston—not the old English Boston, but its big namesake in America. Many parts of the city are indeed quaint and ancient, the finest of the older buildings dating from about the year 1400; but these form only a nucleus for the more modern city which has grown up around them. Coventry now has a population of about seventy-five thousand, and still maintains its old-time reputation as an important manufacturing center. Once it was famed for its silks, ribbons and watches, but this trade was lost to the French and Swiss—some say for lack of a protective tariff. Now cycles and motor cars are the principal products; and we saw several of the famous Daimler cars, made here, being tested on the streets.

Coventry has three fine old churches, whose tall needlelike spires form a landmark from almost any point of view in Warwickshire, and give to the town the appellation by which it is often known—"The City of the Three Spires." Nor could we well have forgotten Coventry's unique legend, for high up on one of the gables of our hotel was a wooden figure said to represent Peeping Tom, who earned eternal ignominy by his curiosity when Lady Godiva resorted to her remarkable expedient to reduce the tax levy of Coventry. Our faith in the story, so beautifully re-told by Tennyson, will not be shaken by the iconoclastic assertion that the effigy is merely an old sign taken from an armourer's shop; that the legend of Lady Godiva is common to half a dozen towns; and that she certainly never had anything to do with Coventry, in any event.

Leaving Coventry the next day about noon in a steady rain, we sought the most direct route to Manchester, thereby missing Nuneaton, the birthplace and for many years the home of George Eliot and the center of some of the most delightful country in Warwickshire. Had we been more familiar with the roads of this country, we could have passed through Nuneaton without loss of time. The distance was only a little greater and over main roads, whereas we traveled for a good portion of the day through narrow byways, and the difficulty of keeping the right road in the continual rain considerably delayed our progress. We were agreeably surprised to find that the car did not skid on the wet macadam road and that despite the rain we could run very comfortably and quite as fast as in fair weather. I had put up our cape top and curtains, but later we learned that it was pleasanter, protected by water-proof wraps, to dash through the rain in the open car. English spring showers are usually light, and it was rather exhilarating to be able to bid defiance to weather conditions that in most parts of the United States would have put a speedy end to our tour.

A few miles farther brought us to Tamworth with its castle, lying on the border between Warwickshire and Staffordshire, the "tower and town" of Scott's "Marmion." The castle of the feudal baron chosen by Scott as the hero of his poem still stands in ruins, and was recently acquired by the town. It occupies a commanding position on a knoll and is surrounded by a group of fine trees.

A dozen miles more over a splendid road brought in view the three spires of Lichfield Cathedral, one of the smallest though most beautiful of these great English churches. Built of red sandstone, rich with sculptures and of graceful and harmonious architecture, there are few cathedrals more pleasing. The town of Lichfield is a comparatively small place, but it has many literary and historical associations, being the birthplace of Dr. Samuel Johnson, whose house is still standing, and for many years the home of Maria Edgeworth. Here, too, once lived Major Andre, whose melancholy death in connection with the American Revolution will be recalled. The cathedral was fortified during the civil war and was sadly battered in sieges by Cromwell's Roundheads; but so completely has it been rebuilt and restored that it presents rather a new appearance as compared with many others. It occurred to us that the hour for luncheon was well past, and we stopped at the rambling old Swan Hotel, which was to all appearances deserted, for we wandered through narrow halls and around the office without finding anyone. I finally ascended two flights of stairs and found a chambermaid, who reluctantly undertook to locate someone in authority, which she at last did. We were shown into a clean, comfortable coffee room, where tea, served in front of a glowing fire place, was grateful indeed after our long ride through the cold rain.

It became apparent that owing to our many delays, we could not easily reach Manchester, and we stopped at Newcastle-under-Lyme for the night. This town has about 20,000 people and lies on the outer edge of the potteries district, where Josiah Wedgewood founded this great industry over one hundred years ago. The whole region comprising Burslem, Hanley, Newcastle, Stoke-on-Trent and many smaller places may be described as a huge, scattered city of about 300,000 inhabitants, nearly all directly or indirectly connected with the manufacture of various grades of china and earthenware. The Castle Hotel, where we stopped, was a very old inn, yet it proved unexpectedly homelike and comfortable. Our little party was given a small private dining room with massive antique furniture, and we were served with an excellent dinner by an obsequious waiter in full-dress suit and with immaculate linen. He cleared the table and left us for the evening with the apartment as a sitting room, and a mahogany desk by the fireside, well supplied with stationery, afforded amends for neglected letters. In the morning, our breakfast was served in the same room, and the bill for entertainment seemed astonishingly low. Mine host will no doubt be wiser in this particular as motorists more and more invade the country.

THE THREE SPIRES OF LICHFIELD.

From Photograph.

An hour's drive brought us to Manchester. The road by which we entered the city took us direct to the Midland Hotel, which is reputed to be the finest in the Kingdom. Manchester is a city of nearly a million inhabitants, but its streets seemed almost like those of a country town as compared with the crowded thoroughfares of London. It is a great center for motoring and I found many of the garages so full that they could not take another car. I eventually came to one of the largest, where by considerable shifting they managed to accommodate my car. But with all this rush of business, it seemed to me that the owners were in no danger of becoming plutocrats, for the charge for a day's garage, cleaning the car, polishing the brass and making a slight repair, was five shillings.

For half the way from Manchester to Leeds, the drive was about as trying as anything I found in England. The road is winding, exceedingly steep in places, and built up on both sides with houses—largely homes of miners and mill operatives. The pavement is of rough cobble-stones, and swarms of dogs and children crowded the way everywhere. Under such conditions, the numerous steep hills, narrow places and sharp turns in the road made progress slow indeed. It was evident that the British motorists generally avoid this country, for we met no cars and our own attracted attention that showed it was not a common spectacle. However, the trip was none the less an interesting one as showing a bit of the country and a phase of English life not usually seen by tourists.

There is little to detain one within the city of Leeds itself, but there are many places of interest in its immediate vicinity. There are few more picturesque spots in Yorkshire than Wharfdale, with its riotous little river and ruins of Bolton Abbey and Barden Tower. This lies about fifteen miles to the northwest, and while for special reasons we went to Ilkley Station by train, the trip is a fine motor drive over good roads. The park which contains the abbey and castle is the property of the Duke of Devonshire, who keeps it at all times open to the public. The River Wharfe, rippling over shingly rocks, leaping in waterfalls and compressed into the remarkable rapids called the Strid, only five or six feet wide but very deep and terribly swift, is the most striking feature of the park. The forest-clad cliffs on either side rise almost precipitously from the edges of the narrow dale, and from their summit, if the climb does not deter one, a splendid view presents itself. The dale gradually opens into a beautiful valley and here the old abbey is charmingly situated on the banks of the river. The ruins are not extensive, but the crumbling walls, bright with ivy and wall flowers, and with the soft green lawn beneath, made a delightful picture in the mottled sunshine and shadows of the English May day.

On our return to Leeds, our friend who accompanied us suggested that we spend the next day, Sunday, at Harrogate, fifteen miles to the north, one of the most famous of English watering places. It had been drizzling fitfully all day, but as we started on the trip, it began to rain in earnest. After picking our way carefully until free from the slippery streets in Leeds, we found the fine macadam road little affected by the deluge. We were decidedly ahead of the season at Harrogate, and there were but few people at the splendid hotel where we stopped.

The following Sunday was as raw and nasty as English weather can be when it wants to, regardless of the time of year, and I did not take the car out of the hotel garage. In the afternoon my friend and I walked to Knaresborough, one of the old Yorkshire towns about three miles distant. I had never even heard of the place before, and it was a thorough surprise to me to find it one of the most ancient and interesting towns in the Kingdom. Not a trace of modern improvement interfered with its old-world quaintness—it looked as if it had been clinging undisturbed to the sharply rising hillside for centuries. Just before entering the town, we followed up the valley of the River Nidd to the so-called "dripping well," whose waters, heavily charged with limestone, drip from the cliffs above and "petrify" various objects in course of time by covering them with a stonelike surface. Then we painfully ascended the hill—not less than a forty-five per cent grade in motor parlance—and wandered through the streets—if such an assortment of narrow foot-paths, twisting around the corners, may be given the courtesy of the name— until we came to the site of the castle. The guide-book gives the usual epitaph for ruined castles, "Dismantled by orders of Cromwell's Parliament," and so well was this done that only one of the original eleven great watch-towers remains, and a small portion of the Norman keep, beneath which are the elaborate vaulted apartments where Becket's murderers once hid. No doubt the great difficulty the Cromwellians had in taking the castle seemed a good reason to them for effectually destroying it. At one time it was in the possession of the notorious Piers Gaveston, and it was for a while the prison-house of King Henry II. There are many other points of interest in Knaresborough, not forgetting the cave from which Mother Shipton issued her famous prophecies, in which she missed it only by bringing the world to an end ahead of schedule time. But they deny in Knaresborough she ever made such a prediction, and prefer to rest her claims to infallibility on her prophecy illustrated on a post card by a highly colored motor car with the legend,

> "Carriages without horses shall go,
> And accidents fill the world with woe."

Altogether, Knaresborough is a town little frequented by Americans, but none the less worthy of a visit. Harrogate is an excellent center for this and many other places, if one is insistent on the very best and most stylish hotel accommodations that the island affords. Ripon, with its cathedral and Fountains Abbey, perhaps the finest ruin in Great Britain, is only a dozen miles away; but we visited these on our return to London from the north.

On Monday the clouds cleared away and the whole country was gloriously bright and fresh after the heavy showers. We returned to Leeds over the road by which we came to Harrogate and which passes Haredale Hall, one of the finest country places in the Kingdom. A large portion of the way the road is bordered by fine forests, which form a great park around the mansion. We passed through Leeds to the southward, having no desire to return to Manchester over the road by which we came, or, in fact, to pass through the city at all. Our objective point for the evening was Chester, and this could be reached quite as easily by passing to the south of Manchester. Wakefield, with its magnificent church, recently dignified as a cathedral, was the first town of consequence on our way, and about twenty-five miles south of Leeds we came to Barnsley, lying on the edge of the great moorlands in central England. There is hardly a town in the whole Kingdom that does not have its peculiar tradition, and an English friend told us that the fame of Barnsley rests on the claim that no hotel in England can equal the mutton chops of the King's Head—a truly unique distinction in a land where the mutton chop is standard and the best in the world.

SUNSET ON THE MOOR.

From Painting by Termohlen.

An English moor is a revelation to an American who has never crossed one and who may have a hazy notion of it from Tennyson's verse or "Lorna Doone." Imagine, lying in the midst of fertile fields and populous cities, a large tract of brown, desolate and broken land, almost devoid of vegetation except gorse and heather, more comparable to the Arizona sagebrush country than anything else, and you have a fair idea of the "dreary, dreary moorland" of the poet. For twenty miles from Barnsley our road ran through this great moor, and, except for two or three wretched-looking public houses—one of them painfully misnamed "The Angel"—there was not a single town or habitation along the road. The moorland road began at Penistone, a desolate-looking little mining town straggling along a single street that dropped down a very sharp grade on leaving the town. Despite the lonely desolation of the moor, the road was excellent, and followed the hills with gentle curves, generally avoiding steep grades. So far as I can recall, we did not meet a single vehicle of any kind in the twenty miles of moorland road—surely a paradise for the scorcher. Coming out of the moor, we found ourselves within half a dozen miles of Manchester—practically in its suburbs, for Stalybridge, Stockport, Altrincham and other large manufacturing towns are almost contiguous with the main city. The streets of these towns were crowded with traffic and streetcar lines are numerous. There is nothing of the slightest interest to the tourist, and after a belated luncheon at a really modern hotel in Stockport, we set out on the last forty miles of our journey. After getting clear of Manchester and the surrounding towns, we came to the Chester road, one of the numberless "Watling Streets," which one finds all over England—a broad, finely kept high way, leading through a delightful country. Northwich, famous for its salt mines, was the only town of any consequence until we reached Chester. We had travelled a distance of about one hundred and twenty miles—our longest day's journey, with one exception—not very swift motoring, but we found that an average of one hundred miles per day was quite enough to thoroughly satisfy us, and even with such an apparently low average as this, a day's rest now and then did not come amiss.

It would be better yet if one's time permitted a still lower daily mileage. Not the least delightful feature of the tour was the marvelous beauty of the English landscapes, and one would have a poor appreciation of these to dash along at forty or even twenty-five miles per hour. There were many places at which we did not stop at all, and which were accorded scant space in the guide-books, that would undoubtedly have given us ideas of English life and closer contact with the real spirit of the people than one could possibly get in the tourist-thronged towns and villages.

V

THE BORDER TOWNS, SHREWSBURY
AND LUDLOW

I shall say but little of Chester, as of every other place on the line of our journey so well known as to be on the itinerary of nearly everybody who makes any pretensions at touring Britain. The volumes which have been written on the town and the many pages accorded it in the guide-books will be quite sufficient for all seekers after information. Frankly, I was somewhat disappointed with Chester. I had imagined its quaintness that of a genuine old country town and was not prepared for the modern city that surrounds its show-places. In the words of an observant English writer: "It seems a trifle self-conscious—its famous old rows carry a suspicion of being swept and garnished for the dollar-distributing visitor from over the Atlantic, and of being less genuine than they really are. However that may be, the moment you are out of these show-streets of Chester, there is a singular lack of charm in the environment. The taint of commerce and the smoke of the north hangs visibly on the horizon. Its immediate surroundings are modern and garish to a degree that by no means assists in the fiction that Chester is the unadulterated old-country town one would like to think it." Such a feeling I could not entirely rid myself of, and even in following the old wall, I could not help noting its carefully maintained disrepair. I would not wish to be understood as intimating that Chester is not well worth a visit, and a visit of several days if one can spare the time; only that its charm was, to me, inferior to that of its more unpretentious neighbors, Shrewsbury and Ludlow. Our stay was only a short one, since our route was to bring us to the town again; still, we spent half a day in a most delightful manner, making a tour of the "rows" and the odd corners with quaint buildings. The tourist, fortified with his red-backed Baedecker, is a common sight to Chester people, and his "dollar-distributing" propensity, as described by the English writer I have quoted, is not unknown even to the smallest fry of the town. Few things during our trip amused me more than the antics of a brown, bare-foot, dirt-begrimed little mite not more than two or three years old, who seized my wife's skirts and hung on for dear life, pouring out earnestly and volubly her unintelligible jargon. We were at first at a loss to understand what our new associate desired, and so grimly did she hang on that it seemed as if another accession to our party was assured—but a light dawned suddenly on us, and, as the brown little hand clasped a broad English copper, our self-appointed companion vanished like a flash into a neighboring shop.

Even when touring in your "wind-shod" car, as an up-to-date English poet puts it, and though your motor waits you not a stone's throw from your hotel, you may not entirely dispense with your antiquated equine friend as a means of locomotion. So we learned when we proposed to visit Eaton Hall, the country place of the Duke of Westminster, which lies closely adjoining Chester, situated deep in the recesses of its eight-thousand-acre park. A conspicuous sign, "Motors strictly forbidden," posted near the great gateway, forced us to have recourse to the hackman, whose moderate charge of eight shillings for a party of three was almost repaid by his services as a guide. He was voluble in his information concerning the Duke and especially dwelt on his distinction as the richest man in the world—an honor which as good and loyal Americans we could not willingly see wrested from our own John D. of oleaginous fame. Eaton Hall is one of the greatest English show-places, but it is modern and might well be matched by the castles of several of our American aristocracy. Tame indeed seemed its swept and garnished newness, its trim and perfect repair, after our visits to so many time-worn places, with their long succession of hoary traditions. The great library, with its thousands of volumes in the richest bindings and its collections of rare editions, might well be the despair of a bibliophile and the pictures and furnishings of rare interest to the connoisseur—but these things one may find in the museums.

Over a main road, almost level and as nearly straight as any English road merits such a description, we covered the forty miles from Chester to Shrewsbury without incident. The most trying grade given in the road-book is one in twenty-five, and all conditions are favorable for record time—in absence of police traps. Four miles out of Chester we passed Rowton Station, lying adjacent to Rowton Moor, where King Charles, standing on the tower of Chester Wall which bears his name, saw his army defeated by the Parliamentarians. We made a late start from Chester, but reached Shrewsbury in time to visit many parts of the town after dinner. We found it indeed a delightful old place, rich in historic traditions, and the center of a country full of interesting places. The town is built on a lofty peninsula, surrounded on three sides by the River Severn, and the main streets lead up exceedingly steep hills. In fact, many of the steepest and most dangerous hills which we found in our travels were in the towns themselves, where grades had been fixed by buildings long ago. The clean macadam in Shrewsbury made it possible to drive our car without chains, though it rained incessantly, but so steep and winding are some of the streets that the greatest caution was necessary.

Shrewsbury is described by an English writer as a "sweet-aired, genuine, dignified and proud old market town, the resort of squires, parsons and farmers, and mainly inhabited by those who minister to their wants. It never dreams of itself as a show-place." He also adds another strong point in its claim to distinction: "Some years ago a book was published by a zealous antiquarian, enumerating with much detail all the families of England of a certain consequence who still occupied either the same estate or estates contiguous to those upon which they were living in the Fifteenth Century. The shire of which Shrewsbury is the capital very easily headed the list in this honorable competition and thereby justified the title of 'proud Salopians,' which the more consequential of its people submit to with much complacency, even though it be not always applied in a wholly serious way."

It is a genuine old border town, so far unspoiled by commercialism. Modern improvements have not invaded its quaint streets to any great extent, and many of these still retain their old names—Dog-pole, Wylecop and Shoplatch—and are bordered by some of the finest half-timbered houses in Britain. Nor is Shrewsbury wanting in famous sons. In front of the old grammar school building is a bronze statue of Charles Darwin, the man who changed the scientific thought of a world, who was born here in 1809. This same grammar school was built in 1630 and is now converted into a museum of Roman relics, which have been found in the immediate vicinity. In its earlier days, many distinguished men received their education here, among them Sir Philip Sidney and Judge Jeffreys. The Elizabethan market-house and the council-house which was visited by both Charles I and James II on different occasions are two of the most fascinating buildings to be seen in the town. There are scant remains, principally of the keep of the castle, built by the Norman baron to whom William the Conqueror generously presented the town. St. Mary is the oldest and most important church, and in some particulars it surpasses the cathedral at Chester. It is architecturally more pleasing and its windows are among the finest examples of antique stained glass in the Kingdom.

We spent some time among the remarkable collection of relics in the museum, and as they mainly came from the Roman city of Uriconium, we planned a side-trip to this place, together with Buildwas Abbey and the old Saxon town of Much Wenlock, all of which are within twenty miles of Shrewsbury. When we left the Raven Hotel it was raining steadily, but this no longer deterred us, and after cautiously descending the steep hill leading out of the town we were soon on the road to Wroxeter, the village lying adjacent to the Roman ruins. We found these of surprising extent and could readily believe the statement made in the local guidebook that a great city was at one time located here. Only a comparatively small portion has been excavated, but the city enclosed by the wall covered nearly one square mile. One great piece of wall about seventy-five feet long and twenty feet in height still stands above ground to mark the place, but the most remarkable revelations were found in the excavations. The foundations of a large public building have been uncovered, and the public baths to which the Romans were so partial are in a remarkable state of preservation, the tile flooring in some cases remaining in its original position. There is every indication that the city was burned and plundered by the wild Welsh tribes sixteen hundred or more years ago.

A few miles farther, mainly through narrow byways, brought us to Buildwas Abbey, beautifully situated near the Severn. Evidently this fine ruin is not much frequented by tourists, for we found no custodian in charge, and the haunts of the old monks had been converted into a sheepfold by a neighboring farmer. Yet at one time it was one of the richest and most extensive monasteries in England. On our return to Shrewsbury, we passed through Much Wenlock, a very ancient town, which also has its ruined abbey. It is remarkable how thickly these monastic institutions were at one time scattered over the Kingdom, and when one considers what such elaborate establishments must have cost to build and to maintain, it is easy to understand why, in the ages of church supremacy, the common people were so miserably poor.

RUINS OF URICONIUM, NEAR SHREWSBURY.

Aside from the places of historic interest that we visited on this trip, the country through which we passed would have made our half day a memorable one. Though the continual rain intercepted the view much of the time, yet from some of the hilltops we had vistas of the Severn Valley with its winding river that we hardly saw surpassed in a country famous for lovely landscapes. We regretted later that our stay at Shrewsbury was so short, for we learned that in the immediate vicinity there are many other places which might well have occupied our attention; but in this case, as in many others, we learned afterwards the things we should have known before our tour began.

Late in the afternoon we started for Ludlow. It was still raining—a gray day with fitful showers that never entirely ceased but only varied in intensity. Much of the beauty of the landscape was hidden in the gray mist, and the distant Welsh hills, rich with soft coloring on clear days, were entirely lost to us. Yet the gloomy day was not altogether without its compensation, for if we had visited Stokesay when the garish sunshine gilded "but to flaunt the ruins gray," we should have lost much of the impression which we retain of the gloom and desolation that so appropriately pervaded the unique old manor with its timbered gatehouse and its odd little church surrounded by thickly set gravestones.

It was only by an accidental glance at our road-book that we saw Stokesay Castle as an "object of interest" on this road about eight miles north of Ludlow. This old house is the finest example in the Kingdom of a fortified manor as distinguished from a castle, its defensive feature being a great crenolated tower, evidently built as a later addition when the manor passed from a well-to-do country gentleman to a member of the nobility. This is actually the case, for there is on record a license granted in 1284 to Lawrence de Ludlow permitting him to "crenolate his house." The house itself was built nearly two hundred years earlier and was later surrounded by a moat as a further means of defense. Considering its age, it is in a wonderfully good state of preservation, the original roof still being intact. We were admitted by the keeper, who lives in the dilapidated but delightfully picturesque half-timbered gatehouse. The most notable feature of the old house is the banqueting hall occupying the greater portion of the first floor, showing how, in the good old days, provision for hospitality took precedence over nearly everything else. Some of the apartments on the second floor retain much of their elaborate oak paneling and there are several fine mantel-pieces. A narrow, circular stairway leads to the tower, from which the beauty of the location is at once apparent. Situated as the mansion is in a lovely valley, bounded by steep and richly wooded hills at whose base the river Onny flows through luxuriant meadows, one is compelled to admire the judgment of the ancient founder who selected the site. It indeed brought us near to the spirit and customs of feudal times as we wandered about in the gloom of the deserted apartments. How comfortless the house must have been—from our standard—even in its best days, with its rough stone floors and rude furnishings! No fireplace appeared in the banqueting hall, which must have been warmed by an open fire, perhaps in the center, as in the hall of Penshurst Place. How little these ancient landmarks were appreciated until recently is shown by the fact that for many years Stokesay Manor was used as a blacksmith-shop and a stable for a neighboring farmer. The present noble proprietor, however, keeps the place in excellent repair and always open to visitors. In one of the rooms of the tower, is exhibited a collection of ancient documents relating to the founding of Stokesay and to its early history.

STOKESAY MANOR HOUSE, NEAR LUDLOW.

After visiting hundreds of historic places during our summer's pilgrimage, the memory of Ludlow, with its quaint, unsullied, old-world air, its magnificent church, whose melodious chime of bells lingers with us yet, its great ruined castle, redolent with romance, and its surrounding country of unmatched interest and beauty, is still the pleasantest of all. I know that the town has been little visited by Americans, and that in Baedeker, that Holy Writ of tourists, it is accorded a scant paragraph in small type. Nevertheless, our deliberately formed opinion is still that if we could re-visit only one of the English towns it would be Ludlow. Mr. A.G. Bradley, in his delightful book, "In the March and Borderland of Wales," which everyone contemplating a tour of Welsh border towns should read, gives an appreciation of Ludlow which I am glad to reiterate when he styles it "the most beautiful and distinguished country town in England." He says: "There are towns of its size perhaps as quaint and boasting as many ancient buildings, but they do not crown an eminence amid really striking scenery, nor yet again share such distinction of type with one of the finest mediaeval castles in England and one possessed of a military and political history unique in the annals of British castles. It is this combination of natural and architectural charm, with its intense historical interest, that gives Ludlow such peculiar fascination. Other great border fortresses were centers of military activities from the Conquest to the Battle of Bosworth, but when Ludlow laid aside its armour and burst out into graceful Tudor architecture, it became in a sense the capital of fourteen counties, and remained so for nearly two hundred years."

We were indeed fortunate in Ludlow, for everything conspired to give us the best appreciation of the town, and were it not for the opinion of such an authority as I have quoted, I might have concluded that our partiality was due to some extent to the circumstances. We had been directed to a hotel by our host in Shrewsbury, but on inquiring of a police officer—they are everywhere in Britain—on our arrival in Ludlow, he did us a great favor by telling us that "The Feathers" hotel just opposite would please us better. We forthwith drew up in front of the finest old black and white building which we saw anywhere in the Kingdom and were given a room whose diamond-paned windows opened toward church and castle. No modern improvements broke in on our old-time surroundings—candles lighted us when the long twilight had faded away.

The splendid dark-oak paneling that reached to the ceiling of the dining room and the richly carved mantel-piece, they told us, were once in rooms of Ludlow Castle. As we sat at our late dinner, a familiar melody from the sonorous chimes of the church-tower came through the open window to our great delight. "O, what a nuisance those bells are," said the neat waiting maid, "and a bad thing for the town, too. Why, the commercials all keep away from Ludlow. They can't sleep for the noise." "Do the chimes ring in the night?" we asked. "At midnight and at four o'clock in the morning," she said, and I was fearful that we would not awake. But we did, and the melody in the silence of the night, amid the surroundings of the quaint old town, awakened a sentiment in us no doubt quite different from that which vexed the soul of the commercial. But we felt that credit was due the honest people of Ludlow, who preferred the music of the sweet-toned bells to sordid business; and, as the maid said, the bells did not awaken anyone who was used to them—surely a fit reward to the citizens for their high-minded disregard of mere material interests.

THE FEATHERS HOTEL, LUDLOW.

I said we were fortunate at Ludlow. The gray, chilly weather and almost continual rain which had followed us for the last few days vanished and the next morning dawned cool and fair, with sky of untainted blue. Our steps were first turned towards the castle, which we soon reached. There was no one to admit us. The custodian's booth was closed, but there was a small gate in the great entrance and we walked in. We had the noble ruin to ourselves, and a place richer in story and more beautiful and majestic in decay we did not find elsewhere. A maze of gray walls rose all around us, but fortunately every part of the ruin bore a printed card telling us just what we wanted to know. The crumbling walls surrounded a beautiful lawn, starred with wild flowers—buttercups and forget-me-nots—and a flock of sheep grazed peacefully in the wide enclosure. We wandered through the deserted, roofless chambers where fireplaces with elaborate stone mantels and odd bits of carving told of the pristine glory of the place. The castle was of great extent, covering the highest point in Ludlow, and before the day of artillery must have been well-nigh impregnable. The walls on the side toward the river rise from a cliff which drops down a sharp incline toward the edge of the water but leaving room for a delightful foot path between rows of fine trees. The stern square tower of the keep, the odd circular chapel with its fine Norman entrance, the great banqueting hall, the elaborate stone fireplaces and the various apartments celebrated in the story of the castle interested us most. From the great tower I saw what I still consider the finest prospect in England, and I had many beautiful views from similar points of vantage. The day was perfectly clear and the wide range of vision covered the fertile valleys and wooded hills interspersed with the villages, the whole country appearing like a vast beautifully kept park. The story of Ludlow Castle is too long to tell here, but no one who delights in the romance of the days of chivalry should fail to familiarize himself with it. The castle was once a royal residence and the two young princes murdered in London Tower by the agents of Richard III dwelt here for many years. In 1636 Milton's "Mask of Comus," suggested by the youthful adventures of the children of the Lord President, was performed in the castle courtyard. The Lord of the castle at one time was Henry Sidney, father of Sir Philip, and his coat-of-arms still remains over one of the entrances. But the story of love and treason, of how in the absence of the owner of the castle, Maid Marion admitted her clandestine lover, who brought a hundred armed men at his back to slay the inmates and capture the fortress, is the saddest and most tragic of all. We saw high up in the wall, frowning over the river, the window of the chamber from which she had thrown herself after slaying her recreant lover in her rage and despair. A weird story it is, but if the luckless maiden still haunts the scene of her blighted love, an observant sojourner who fitly writes of Ludlow in poetic phrase never saw her. "Nearly every midnight for a month," he says, "it fell to me to traverse the quarter of a mile of dark, lonely lane that leads beneath the walls of the castle to the falls of the river, and a spot more calculated to invite the wanderings of a despairing and guilty spirit, I never saw. But though the savage gray towers far above shone betimes in the moonlight and the tall trees below rustled weirdly in the night breeze and the rush of the river over the weir rose and fell as is the wont of falling water in the silence of the night, I looked in vain for the wraith of the hapless maiden of the heath and finally gave up the quest."

LUDLOW CASTLE, THE KEEP AND ENTRANCE.

When we left the castle, though nearly noon, the custodian was still belated, and we yet owe him sixpence for admittance, which we hope to pay some time in person. A short walk brought us to the church—"the finest parish church in England," declares one well qualified to judge. "Next to the castle," he says, "the glory of Ludlow is its church, which has not only the advantage of a commanding site but, as already mentioned, is held to be one of the finest in the country." It is built of red sandstone and is cruciform in shape, with a lofty and graceful tower, which is a landmark over miles of country and beautiful from any point of view. I have already mentioned the chime of bells which flings its melodies every few hours over the town and which are hung in this tower. The monuments, the stained-glass windows and the imposing architecture are scarcely equalled by any other church outside of the cathedrals.

We had made the most of our stay in Ludlow, but it was all too short. The old town was a revelation to us, as it would be to thousands of our countrymen who never think of including it in their itinerary. But for the motor car, it would have remained undiscovered to us. With the great growth of this method of touring, doubtless thousands of others will visit the place in the same manner, and be no less pleased than we were.

From Ludlow we had a fine run to Worcester, though the road was sprinkled with short, steep hills noted "dangerous" in the road-book. Our fine weather was very transient, for it was raining again when we reached Worcester. We first directed our steps to the cathedral, but when nearly there beheld a large sign, "This way to the Royal Porcelain Works," and the cathedral was forgotten for the time by at least one member of our party. The Royal Porcelain Works it was, then, for hadn't we known of Royal Worcester long before we knew there was any cathedral—or any town, for that matter? It is easy to get to the Royal Porcelain Works: a huge sign every block will keep you from going astray and an intelligent guide will show you every detail of the great establishment for only a sixpence. But it is much harder and more costly to get away from the Royal Worcester Works, and when we finally did we were several guineas poorer and were loaded with a box of fragile ware to excite the suspicions of our amiable customs officials. Nevertheless, the visit was full of interest. Our guide took us through the great plant from the very beginning, showing us the raw materials—clay, chalk and bones—which are ground to a fine powder, mixed to a paste, and deftly turned into a thousand shapes by the skilled potter. We were shown how the bowl or vase was burned, shrinking to nearly half its size in the process. We followed the various steps of manufacture until the finished ware, hand-painted, and burned many times to bring out the colors, was ready for shipment. An extensive museum connected with the works is filled with rare specimens to delight the soul of the admirer of the keramic art. There were samples of the notable sets of tableware manufactured for nearly every one of the crowned heads of Europe during the last century, gorgeous vases of fabulous value, and rare and curious pieces without number.

When we left the porcelain works it was too late to get into the cathedral, and when we were ready to start in the morning it was too early. So we contented ourselves with driving the car around the noble pile and viewing the exterior from every angle. We took the word of honest Baedeker that the interior is one of the most elaborate and artistic in England but largely the result of modern restoration. The cathedral contains the tomb of King John, who requested that he be buried here, though his life was certainly not such as to merit the distinction. Here, too, is buried the elder brother of King Henry VIII, Prince Arthur, who died at Ludlow Castle in 1502; and had he lived to be king in place of the strenuous Henry, who can say what changes might have been recorded in English history? All these we missed; nor did we satisfy ourselves personally of the correctness of the claim that the original entry of the marriage contract of William Shakespeare and Anne Hathaway is on file in the diocese office near the gateway of the cathedral. Along with the other notable places of the town mentioned in the guide-book as worthy of a visit is the great factory where the fiery Worcestershire sauce is concocted, but this did not appeal to our imagination as did the porcelain works. Our early start and the fine, nearly level road brought us to Stratford-upon-Avon well before noon. Here we did little more than re-visit the shrines of Shakespeare—the church, the birthplace, the grammar school—all familiar to the English-speaking world. Nor did we forget the Red Horse Inn at luncheon time, finding it much less crowded than on our previous visit, for we were still well in advance of the tourist season. After luncheon we were lured into a shop across the street by the broad assurance made on an exceedingly conspicuous sign that it is the "largest souvenir store on earth." Here we hoped to secure a few mementos of our visit to Stratford by motor car. We fell into a conversation with the proprietor, a genial, white-haired old gentleman, who, we learned, had been Mayor of the town for many years—and is it not a rare distinction to be Mayor of Shakespeare's Stratford? The old gentleman bore his honors lightly indeed, for he said he had insistently declined the office but the people wouldn't take no for an answer.

It is only a few miles to Warwick over winding roads as beautiful as any in England. One of these leads past Charlecote, famous for Shakespeare's deer-stealing episode, but no longer open to the public. We passed through Warwick—which reminded us of Ludlow except for the former's magnificent situation—without pausing, a thing which no one would do who had not visited that quaint old town some time before. In Leamington, three miles farther on, we found a modern city of forty thousand inhabitants, noted as a resort and full of pretentious hotels. After we were located at the Manor House there was still time for a drive to Kenilworth Castle, five miles away, to which a second visit was even more delightful than our previous one. For the next day we had planned a circular tour of Warwickshire, but a driving, all-day rain and, still more, the indisposition of one of our party, confined us to our hotel. Our disappointment was considerable, for within easy reach of Leamington there were many places that we had planned to visit. Ashow Church, Stoneleigh Abbey, George Eliot's birthplace and home near Nuneaton, the cottage of Mary Arden, mother of Shakespeare, Rugby, with its famous school, and Maxstoke Castle—an extensive and picturesque ruin—are all within a few miles of Leamington.

From Leamington to London was nearly an all-day's run, although the distance is only one hundred miles. A repair to the car delayed us and we went several miles astray on the road. It would have been easier to have returned over the Holyhead Road, but our desire to see more of the country led us to take a route nearly parallel to this, averaging about fifteen miles to the southward. Much of the way this ran through narrow byways and the country generally lacked interest. We passed through Banbury, whose cross, famous in nursery rhyme, is only modern. At Waddesdon we saw the most up-to-date and best ordered village we came across in England, with a fine new hotel, the Five Arrows, glittering in fresh paint. We learned that this village was built and practically owned by Baron Rothschild, and just adjoining it was the estate which he had laid out. The gentleman of whom we inquired courteously offered to take us into the great park, and we learned that he was the head landscape gardener. The palace is modern, of Gothic architecture, and crowns an eminence in the park. It contains a picture gallery, with examples of the works of many great masters, which is open to the public on stated days of the week.

On reaching London, we found that our tour of the Midlands had covered a little less than eight hundred miles, which shows how much that distance means in Britain when measured in places of historic and literary importance, of which we really visited only a few of those directly on the route of our journey or lying easily adjacent to it.

VI

LONDON TO LAND'S END

The road from London to Southampton is one of the oldest in the Kingdom and passes many places of historic interest. In early days this highway, leading from one of the main seaports through the ancient Saxon capital, was of great importance. Over this road we began the trip suggested by the Touring Secretary of the Motor Union. As usual, we were late in getting started and it was well after noon when we were clear of the city. At Kingston-on-Thames, practically a suburb, filled with villas of wealthy Londoners, we stopped for lunch at the Griffin Hotel, a fine old inn whose antiquity was not considered sufficient to atone for bad service, which was sometimes the case. Kingston has a history as ancient as that of the capital itself. Its name is peculiar in that it was not derived from King's Town, but from King's Stone; and at the town crossing is the identical stone, so says tradition, upon which the Saxon kings were crowned. It would seem to one that this historic bit of rock would form a more fitting pedestal for the English coronation chair than the old Scottish stone from Dunstafnage Castle.

After a short run from Kingston, we passed down High Street, Guildford, which, a well qualified authority declares, is "one of the most picturesque streets in England." Guildford might well detain for a day or more anyone whose time will permit him to travel more leisurely than ours did. William Cobbett, the author and philosopher, who was born and lived many years near by, declared it "the happiest looking town he ever knew"—just why, I do not know. The street with the huge town clock projecting half way across on one side, the Seventeenth Century Town Hall with its massive Greek portico on the other, and a queerly assorted row of many-gabled buildings following its winding way, looked odd enough, but as to Guildford's happiness, a closer acquaintance would be necessary.

Shortly after leaving the town, the ascent of a two-mile hill brought us to a stretch of upland road which ran for several miles along a tableland lying between pleasantly diversified valleys sloping on either side. From this a long, gradual descent led directly into Farnham, the native town of William Cobbett. The house where he was born and lived as a boy is still standing as "The Jolly Farmers' Inn." One may see the little house which was the birthplace of the Rev. Augustus Toplady, whose hymn, "Rock of Ages," has gained world-wide fame. On the hill overlooking the town is the ancient castle, rebuilt in the Sixteenth Century and from that time one of the palaces of the bishops of Winchester. Here, too, lingers one of the ubiquitous traditions of King Charles I, who stopped at Vernon House in West Street while a prisoner in the hands of the Parliamentarians on their way to

London. A silk cap which the king presented to his host is proudly shown by one of the latter's descendants, who is now owner of the house.

One must be well posted on his route when touring Britain or he will pass many things of note in sublime ignorance of their existence. Even the road-book is not an infallible guide, for we first knew that we were passing through Chawton when the postoffice sign, on the main street of a straggling village, arrested our attention. We were thus reminded that in this quiet little place the inimitable Jane Austin had lived and produced her most notable novels, which are far more appreciated now than in the lifetime of the authoress. An old woman of whom we inquired pointed out the house—a large square building with tiled roof, now used as the home of a workingmen's club. Less than two miles from Chawton, though not on the Winchester road, is Selborne, the home of Gilbert White, the naturalist, and famed as one of the quaintest and most retired villages in Hampshire.

But one would linger long on the way if he paused at every landmark on the Southampton road. We had already loitered in the short distance which we had traveled until it was growing late, and with open throttle our car rapidly covered the last twenty miles of the fine road leading into Winchester.

From an historical point of view, no town in the Kingdom surpasses the proud old city of Winchester. The Saxon capital still remembers her ancient splendor and it was with a manifest touch of pride that the old verger who guided us through the cathedral dwelt on the long line of kings who had reigned at Winchester before the Norman conquest. To him, London at best was only an upstart and an usurper. Why,

> "When Oxford was shambles
> And Westminster was brambles,
> Winchester was in her glory."

And her glory has never departed from her and never will so long as her great cathedral stands intact, guarding its age-long line of proud traditions. The exterior is not altogether pleasing—the length exceeding that of any cathedral in Europe, together with the abbreviated tower, impresses one with a painful sense of lack of completeness and a failure of proper proportion. It has not the splendid site of Durham or Lincoln, the majesty of the massive tower of Canterbury, or the grace of the great spire of Salisbury. But its interior makes full amends. No cathedral in all England can approach it in elaborate carvings and furnishings or in interesting relics and memorials. Here lie the bones of the Saxon King Ethelwulf, father of Alfred the Great; of Canute, whose sturdy common sense silenced his flatterers; and of many others. A scion of the usurping Norman sleeps here too, in the tomb where William Rufus was buried, "with many looking on and few grieving." In the north aisle a memorial stone covers the grave of Jane Austen and a great window to her memory sends its many-colored shafts of light from above. In the south transept rests Ike Walton, prince of fishermen, who, it would seem to us, must have slept more peacefully by some rippling brook. During the Parliamentary wars Winchester was a storm center and the cathedral suffered severely at the hands of the Parliamentarians. Yet fortunately, many of its ancient monuments and furnishings escaped the wrath of the Roundhead iconoclasts. The cathedral is one of the oldest in England, having been mainly built in the Ninth Century. Recently it has been discovered that the foundations are giving away to an extent that makes extensive restoration necessary, but it will be only restored and not altered in any way.

But we may not pause long to tell the story of even Winchester Cathedral in this hasty record of a motor flight through Britain. And, speaking of the motor car, ardent devotee as I am, I could not help feeling a painful sense of the inappropriateness of its presence in Winchester; of its rush through the streets at all hours of the night; of its clatter as it climbed the steep hills in the town; of the blast of its unmusical horn; and of its glaring lights, falling weirdly on the old buildings. It seemed an intruder in the capital of King Alfred.

There is much else in Winchester, though the cathedral and its associations may overshadow everything. The college, one of the earliest educational institutions in the Kingdom, was founded about 1300, and many of the original buildings stand almost unchanged. The abbey has vanished, though the grounds still serve as a public garden; and of Wolvesley Palace, a castle built in 1138, only the keep still stands. How usual this saying, "Only the keep still stands," becomes of English castles,—thanks to the old builders who made the keep strong and high to withstand time, and so difficult to tear down that it escaped the looters of the ages.

A day might well be given to the vicinity of Winchester, which teems with points of literary and historic interest. In any event, one should visit Twyford, only three miles away, often known as the "queen of the Hampshire villages" and famous for the finest yew tree in England. It is of especial interest to Americans, since Benjamin Franklin wrote his autobiography here while a guest of Dr. Shipley, Vicar of St. Asaph, whose house, a fine Elizabethan mansion, still stands.

To Salisbury by way of Romsey is a fine drive of about thirty miles over good roads and through a very pleasing country. Long before we reached the town there rose into view its great cathedral spire, the loftiest and most graceful in Britain, a striking landmark from the country for miles around. Following the winding road and passing through the narrow gateway entering High Street, we came directly upon this magnificent church, certainly the most harmonious in design of any in the Kingdom. The situation, too, is unique, the cathedral standing entirely separate from any other building, its gray walls and buttresses rising sheer up from velvety turf such as is seen in England alone. It was planned and completed within the space of fifty years, which accounts for its uniformity of style; while the construction of most of the cathedrals ran through the centuries with various architecture in vogue at different periods. The interior, however, lacks interest, and the absence of stained glass gives an air of coldness. It seems almost unbelievable that the original stained windows were deliberately destroyed at the end of the Eighteenth Century by a so-called architect, James Wyatt, who had the restoration of the cathedral in charge. To his everlasting infamy, "Wyatt swept away screens, chapels and porches, desecrated and destroyed the tombs of warriors and prelates, obliterated ancient paintings; flung stained glass by cart loads into the city ditch; and razed to the ground the beautiful old campanile which stood opposite the north porch." That such desecration should be permitted in a civilized country only a century ago indeed seems incredible.

A COTTAGE IN HOLDENHURST, HAMPSHIRE.

From Water Color by Noelsmith.

A GLADE IN NEW FOREST.

No one who visits Salisbury will forget Stonehenge, the most remarkable relic of prehistoric man to be found in Britain. Nearly everyone is familiar with pictures of this solitary circle of stones standing on an eminence of Salisbury Plain, but one who has not stood in the shadow of these gigantic monoliths can have no idea of their rugged grandeur. Their mystery is deeper than that of Egypt's sphynx, for we know something of early Egyptian history, but the very memory of the men who reared the stones on Salisbury Plain is forgotten. Who they were, why they built this strange temple, or how they brought for long distances these massive rocks that would tax modern resources to transport, we have scarcely a hint. The stones stand in two concentric circles, those of the inner ring being about half the height of the outer ones. Some of the stones are more than twenty feet high and extend several feet into the ground. There are certain signs which seem to indicate that Stonehenge was the temple of some early sun-worshiping race, and Sir Norman Lockyer, who has made a special study of the subject, places the date of construction about 1680 B.C. No similar stone is found in the vicinity; hence it is proof positive that the builders of Stonehenge must have transported the enormous monoliths for many miles. The place lies about eight miles north of Salisbury. We went over a rather lonely and uninteresting road by the way of Amesbury, which is two miles from Stonehenge. We returned by a more picturesque route, following the River Avon to Salisbury and passing through Millston, a quaint little village where Joseph Addison was born in 1672.

A few miles south of Salisbury we entered New Forest, an ancient royal hunting domain covering nearly three hundred square miles and containing much of the most pleasing woodland scenery in England. This is extremely diversified but always beautiful. Glades and reaches of gentle park and meadow and open, heath-like stretches contrast wonderfully with the dark masses of huge oaks and beeches, under some of which daylight never penetrates. We stopped for the night at Lyndhurst, directly in the center of the forest and sometimes called the capital of New Forest. It looks strangely new for an English town, and the large church, built of red brick and white stone, shows its recent origin. In this church is a remarkable altar fresco which was executed by the late Lord Leighton. The fine roads and splendid scenery might occupy at least a day if time permitted; but if, like us, one must hasten onward, a run over the main roads of New Forest will give opportunity to see much of its sylvan beauty.

Our route next day through the narrow byways of Dorsetshire was a meandering one. From Lyndhurst we passed through Christchurch, Blandford and Dorchester and came for the night to Yeovil. We passed through no place of especial note, but no day of our tour afforded us a better idea of the more retired rural sections of England. By the roadside everywhere were the thatched roof cottages with their flower gardens, and here and there was an ancient village which to all appearances might have been standing quite the same when the Conqueror landed in Britain. Oftentimes the byways were wide enough for only one vehicle, but were slightly broadened in places to afford opportunity for passing. Many of the crossings lacked the familiar sign-boards, and the winding byways, with nothing but the map for a guide, were often confusing, and sharp turns between high hedges made careful driving necessary. At times we passed between avenues of tall trees and again unexpectedly dropped into some quiet village nestling in the Dorset hills. One of the quaintest of these, not even mentioned in Baedeker, is Cerne Abbas, a straggling village through which the road twisted along—a little old-world community, seemingly severed from modern conditions by centuries. It rather lacked the cozy picturesqueness of many English villages. It seemed to us that it wanted much of the bloom and shrubbery. Everywhere were the gray stone houses with thatched roofs, sagging walls and odd little windows with square or diamond-shaped panes set in iron casements. Nowhere was there a structure that had the slightest taint of newness. The place is quite unique. I do not recall another village that impressed us in just the same way. Our car seemed strangely out of place as it cautiously followed the crooked main street of the town, and the attention bestowed on it by the smaller natives indicated that a motor was not a common sight in Cerne Abbas. Indeed, we should have missed it ourselves had we not wandered from the main road into a narrow lane that led to the village. While we much enjoyed our day in the Dorset byways, our progress had necessarily been slow.

In Yeovil, we found an old English town apparently without any important history, but a prosperous center for a rich farming country. The place is neat and clean and has a beautifully kept public park—a feature of which the average English town appears more appreciative than the small American city.

From Yeovil to Torquay, through Exeter, with a stop at the latter place, was an unusually good day's run. The road was more hilly than any we had passed over heretofore, not a few of the grades being styled "dangerous," and we had been warned by an English friend that we should find difficult roads and steep hills in Devon and Cornwall. However, to one who had driven over some of our worst American roads, even the "bad" roads of England looked good, and the "dangerous" hills, with their smooth surface and generally uniform grade, were easy for our moderate-powered motor.

Exeter enjoys the distinction of having continuously been the site of a town or city for a longer period than is recorded of any other place in England. During the Roman occupation it was known as a city, and it is believed that the streets, which are more regular than usual and which generally cross each other at right angles, were first laid out by the Romans. It is an important town of about fifty thousand inhabitants, with thriving trade and manufactures, and modern improvements are in evidence everywhere.

The cathedral, though not one of the largest or most imposing, is remarkable for the elaborate carving of the exterior. The west front is literally covered with life-sized statues set in niches in the wall, but the figures are all sadly time-worn, many of them having almost crumbled away. Evidently the Roundheads were considerate of Exeter Cathedral that such a host of effigies escaped destruction at their hands; and they were not very well disposed towards Exeter, either, as it was always a Royalist stronghold. Possibly it was spared because the Cromwellians found it useful as a place of worship, and in order to obtain peace and harmony between the two factions of the army the cathedral was divided into two portions by a high brick wall through the center, the Independents holding forth on one side and the Presbyterians on the other.

The road from Exeter to Torquay follows the coast for some distance, affording many fine views of the ocean. We were now in the "limestone country," and the roads are exceedingly dusty in dry weather. The dust, in the form of a fine white powder, covers the trees and vegetation, giving the country here and there an almost ghostly appearance. No wonder that in this particular section there is considerable prejudice against the motor on account of its great propensity to stir up the dust. So far as we ourselves were concerned, we usually left it behind us, and it troubled us only when some other car got in ahead of us.

Torquay is England's Palm Beach—a seacoast-resort town where the temperature rarely falls below forty degrees, thanks to the warm current of the Gulf Stream; and where the sea breezes keep down the summer heat, which seldom rises above sixty degrees. It is especially a winter resort, although the hotels keep open during the year. Most of the town is finely situated on a high promontory overlooking a beautiful harbor, studded with islands and detached rocks that half remind one of Capri. From our hotel window we had a glorious ocean view, made the more interesting for the time being by a dozen of King Edward's men-of-war, supposed to be defending Torquay against "the enemy" of a mimic naval warfare.

On the opposite side of Tor Bay is the quiet little fishing village of Brixham, the landing-place of Prince William of Orange. We reached here early on a fine June day when everything was fresh after heavy showers during the night. The houses rise in terraces up the sharp hillside fronting the harbor, which was literally a forest of fishing-boat masts. A rather crude stone statue of William stands on the quay and a brass foot-print on the shore marks the exact spot where the Dutch prince first set foot in England, accompanied by an army of thirteen thousand men. Our car attracted a number of urchins, who crowded around it and, though we left it unguarded for an hour or more to go out on the sea-wall and look about the town, not one of the fisher lads ventured to touch it or to molest anything—an instance of the law-abiding spirit which we found everywhere in England.

From Brixham, an hour's drive over bad roads brought us to Dartmouth, whither we had been attracted by the enthusiastic language of an English writer who asserts that "There is scarcely a more romantic spot in the whole of England than Dartmouth. Spread out on one of the steep slopes of the Dart, it overlooks the deep-set river toward the sea. Steep wooded banks rising out of the water's edge give the winding of the estuaries a solemn mystery which is wanting in meadows and plough-land. In the midst of scenery of this character—and it must have been richer still a few centuries back—the inhabitants of Dartmouth made its history."

As we approached the town, the road continually grew worse until it was little better than the average unimproved country highway in America, and the sharp loose stones everywhere were ruinous on tires. It finally plunged sharply down to a steamboat ferry, over which we crossed the Dart and landed directly in the town. There are few towns in England more charmingly located than old Dartmouth, and a hundred years ago it was an important seaport, dividing honors about equally with Plymouth.

The road to Dartmouth was unusually trying; the route which we took to Plymouth was by odds the worst of equal distance we found anywhere. We began with a precipitous climb out of the town, up a very steep hill over a mile long, with many sharp turns that made the ascent all the more difficult. We were speedily lost in a network of unmarked byways running through a distressingly poor-looking and apparently quite thinly inhabited country. After a deal of studying the map and the infrequent sign-boards we brought up in a desolate-looking little village, merely a row of gray stone, slate-roofed houses on either side of the way, and devoid of a single touch of the picturesque which so often atones for the poverty of the English cottages. No plot of shrubbery or flower-garden broke the gray monotony of the place. We had seen nothing just like it in England, though some of the Scotch villages which we saw later, matched it very well.

Here a native gave us the cheerful information that we had come over the very road we should not have taken; that just ahead of us was a hill where the infrequent motor cars generally stalled, but he thought that a good strong car could make it all right. Our car tackled the hill bravely enough, but slowed to a stop before reaching the summit; but by unloading everybody except the driver, and with more or less coaxing and adjusting, it was induced to try it again, with a rush that carried it through. The grade, though very steep, was not so much of an obstacle as the deep sand, with which the road was covered. We encountered many steep hills and passed villages nearly as unprepossessing as the first one before we came to the main Plymouth-Exeter road, as excellent a highway as one could wish. It was over this that our route had originally been outlined, but our spirit of adventure led us into the digression I have tried to describe. It was trying at the time, but we saw a phase of England that we otherwise would have missed and have no regrets for the strenuous day in the Devonshire byways.

Plymouth, with the adjoining towns of Devonport and Stonehouse, is one of the most important seaports in the Kingdom, the combined population being about two hundred thousand. The harbor is one of the best and affords safe anchorage for the largest ocean-going vessels. It is protected by a stupendous granite breakwater, costing many millions and affording a delightful promenade on a fine day. Plymouth is the principal government naval port and its ocean commerce is gaining rapidly on that of Liverpool. To Americans it appeals chiefly on account of its connection with the Pilgrim Fathers, who sailed from its harbor on the Mayflower in 1620. A granite block set in the pier near the oldest part of the city is supposed to mark the exact spot of departure of the gallant little ship on the hazardous voyage, whose momentous outcome was not then dreamed of. I could not help thinking what a fine opportunity is offered here for some patriotic American millionaire to erect a suitable memorial to commemorate the sailing of the little ship, fraught with its wonderful destiny. The half day spent about the old city was full of interest; but the places which we missed would make a most discouraging list. It made us feel that one ought to have two or three years to explore Britain instead of a single summer's vacation.

ROCKS OFF CORNWALL.

From Painting by Warne Browne. Exhibited 1906 Royal Academy.

From Plymouth to Penzance through Truro runs the finest road in Cornwall, broad, well kept and with few steep grades. It passes through a beautiful section and is bordered in many places by the immense parks of country estates. In some of these the woods were seemingly left in their natural wild state, though close inspection showed how carefully this appearance was maintained by judicious landscape gardening. In many of the parks, the rhododendrons were in full bloom, and their rich masses of color wonderfully enlivened the scenery. Everything was fresh and bright. It had been raining heavily the night before and the air was free from the dust that had previously annoyed us. It would be hard to imagine anything more inspiring than the vistas which opened to us as we sped along. The road usually followed the hills in gentle curves, but at places it rose to splendid points of vantage from which to view the delightful valleys. Then again it lost itself under great over-arching trees, and as we came too rapidly down a steep hill on entering Bodmin, the road was so heavily shaded that we were near our undoing. The loose sand had been piled up by the rain and the dense shade prevented the road from drying. The car took a frightful skid and by a mere hair's breadth escaped disastrous collision with a stone wall—but we learned something.

After leaving Truro, an ancient town with a recently established cathedral, the road to Penzance, though excellent, is without special interest. It passes through the copper-mining section of Cornwall and the country is dotted with abandoned mines. A few are still operated, but it has come to the point where, as a certain Englishman has said, "Cornwall must go to Nevada for her copper," and there are more Cornish miners in the western states than there are in their native shire.

Penzance is another of the South of England resort towns and is beautifully situated on Mounts Bay. One indeed wonders at the great number of seacoast resorts in Britain, but we must remember that there are forty millions of people in the Kingdom who need breathing places as well as a number of Americans who come to these resorts. The hotels at these places are generally excellent from the English point of view, which differs somewhat from the American. Probably there is no one point on which the difference is greater than the precise temperature that constitutes personal comfort and makes a fire in the room necessary. On a chilly, muggy day when an American shivers and calls for a fire in the generally diminutive grate in his room, the native enjoys himself or even complains of the heat, and is astonished at his thin-skinned cousin, who must have his room—according to the British notion—heated to suffocation. The hotel manager always makes a very adequate charge for fires in guest-rooms and is generally chary about warming the corridors or public parts of the hotel. In one of the large London hotels which actually boasts of steam heat in the hallways, we were amazed on a chilly May day to find the pipes warm and a fine fire blazing in the great fireplace in the lobby. The chambermaid explained the astonishing phenomenon: the week before several Americans had complained frequently of the frigid atmosphere of the place without exciting much sympathy from the management, but after they had left the hotel, it was taken as an evidence of good faith and the heat was turned on. But this digression has taken me so far away from Penzance that I may as well close this chapter with it.

VII

FROM CORNWALL TO SOUTH WALES

In following a five-thousand-mile motor journey through Britain, there will be little to say of Penzance, a pleasant resort town, yet without anything of notable importance. A mile farther down the coast is Newlyn, a fishing-village which has become a noted resort for artists and has given its name to a school of modern painting. A handsome building for a gallery and art institute, and which also serves as headquarters for the artists, has recently been erected by a wealthy benefactor. We walked over to the village, hoping to learn that the fisher-fleet would be in the next morning, but were disappointed. A man of whom we inquired informed us that the fishermen would not bring in their catch until two days later. He seemed to recognize at once that we were strangers—Americans, they all know it intuitively—and left his task to show us about the immense quay where the fishermen dispose of their catch at auction. He conducted us out on the granite wall, built by the Government to enclose the harbor and insuring the safety of the fisher-fleet in fiercest storms. He had been a deep-sea fisherman himself and told us much of the life of these sturdy fellows and the hardships they endure for little pay.

The ordinary fishing boat is manned by five or six men and makes two trips each week to the deep-sea fishing "grounds," seventy-five to one hundred miles away. The craft is rude and comfortless in the extreme and so constructed as to be nearly unsinkable if kept off the rocks. The fish are taken by trawling great nets and drawing them aboard with a special tackle. The principal catch of the New-lyn fishermen is herring, which are pickled in the village and exported, mainly to Norway and Sweden. The value of the fish depends on the state of the market, and the price realized is often as low as a shilling per hundred weight. The majority of the population of Cornwall is engaged directly or indirectly in the fisheries, and considering the inferiority of most of the country for agriculture and the extensive coast line with its numerous harbors, it is not strange that so many of the natives should follow this life. In earlier days, smuggling and wrecking constituted the occupation of a large number of the Cornishmen, but under modern conditions these gentle arts can no longer be successfully practiced, and fishing furnishes about the only alternative.

NEAR LAND'S END.

From Water Color by Wm. T. Richards.

Just across the peninsula is St. Ives, another fishing village, even more pictur-esque than Newlyn and quite as much in favor with the artists. To reach this town we turned a few miles from the main road on the following day, but missed the fisher-fleet as before. The bay on which St. Ives is situated is the most beautiful on the Cornish coast, and on the day of our visit the bright stretch of water, sleeping placidly under the June skies and dotted with glistening sails, well maintained its reputation for surpassing loveliness. Before we entered the town a man of whom we inquired the way advised us to leave our car and walk down the sharp descent to the coast, where the village mostly lies. The idea of the return trip was not pleasing, and we boldly started down, only to wish we had been more amenable to the friendly advice, for a steeper, narrower, crookeder street we did not find anywhere. In places it was too narrow for vehicles to pass abreast, and sharp turns on a very steep grade, in streets crowded with children, made the descent ex-ceedingly trying. However, we managed to get through safely and came to a stop directly in front of the Fifteenth Century church, an astonishingly imposing struc-ture for a village which showed more evidences of poverty than of anything else. The church was built at a time when the smugglers and wreckers of Cornwall no doubt enjoyed greater prosperity and felt, perhaps, more anxiety for their souls' welfare than do their fisher-folk descendants.

On re-ascending the hill we stopped at the Castle for our noonday luncheon, but the castle in this instance is a fine old mansion built about a hundred years ago as a private residence and since passed into the possession of a railway company, which has converted it into an excellent hotel. Situated as it is, in a fine park on the eminence overlooking the bay, few hostelries at which we paused seemed more inviting for a longer sojourn.

Four miles from Penzance is Marazion, and St. Michael's Mount, lying near at hand, takes its name from the similar but larger and more imposing cathe-dral-crowned headland off the coast of France. It is a remarkable granite rock, con-nected with the mainland by a strip of sand, which is clear of the water only four hours of the day. The rock towers to a height of two hundred and fifty feet and is about a mile in circumference. It is not strange that in the days of castle-building such an isolated site should have been seized upon; and on the summit is a ma-ny-towered structure built of granite and so carefully adapted to its location as to seem almost a part of the rock itself. When we reached Marazion, the receding tide had left the causeway dry, and as we walked leisurely the mile or so between the town and the mount, the water was already stealthily encroaching on the path-way. We found the castle more of a gentleman's residence than a fortress, and it was evidently never intended for defensive purposes. It has been the residence of the St. Aubyn family since the time of Charles II, and the villagers were all agog over elaborate preparations to celebrate the golden wedding anniversary of the present proprietor. The climb is a wearisome one, and we saw little of the castle, being admitted only to the entrance-hall and the small Gothic chapel, which was undergoing restoration; but the fine view from the battlements alone is worth the effort. The castle never figured in history and is remarkable chiefly for its unique location. By the time of our return the tide had already risen several feet and we were rowed to the mainland in a boat.

ON DARTMOOR.

From Water Color by Vincent.

On our return to Truro we took the road by which we came, but on leaving there our road roughly followed the Northern Cornish coast, and at intervals we caught glimpses of the ocean. For some distance we ran through a rough moorland country, although the road was comparatively level and straight. We passed Camelford—which some say is the Camelot of the Arthur legends—only five miles distant from the ruins of Tintagel Castle on the coast, and came early to Launceston, where the clean hospitable-looking White Hart Hotel offered strong inducements to stop for the night. A certain weariness of the flesh, resulting from our run over the last long stretch of the moorland road, was an equally important factor in influencing our action.

Launceston was one of the surprises that we frequently came across—a town that we had never heard of before and doubtless one that an American seldom sees. Yet the massive castle, whose circular keep crowns an eminence overlooking the town, was one of the objects that loomed into view long before we reached the place, and its gloomy grandeur, as we wandered through its ruins in the fading twilight, deeply impressed us. A rude stairway led to the top of the great circular tower, rising high above the summit of the hill, which itself dominates the country, and the view stretching away in every direction was far-reaching and varied. The castle has been gradually falling into ruin for the last six hundred years, but in its palmy days it must have been one of the grimmest and most awe-inspiring of the fortresses in the west country. Scarcely another ruin did we see anywhere more imposing in location and more picturesque in decay. Masses of ivy clung to the crumbling walls and all around spread a beautiful park, with soft, velvety turf interspersed with shrubbery and bright dashes of color from numerous well cared-for flower beds.

Not less unique is St. Steven's church, the like of which is not to be found elsewhere in Britain. Its walls are covered with a network of fine carving, vine and flower running riot in stone, and they told us that this was done by English stonecutters, though nearly all such carving on the cathedrals was the work of artisans from the continent. The Launceston church is pointed to as an evidence that English workmen could have done quite as well had they been given the chance. Aside from this wonderful carving, which covers almost every stone of the exterior, the church is an imposing one and has lately been restored to its pristine magnificence. Launceston had its abbey, too, but this has long since disappeared, and all that now remains of it is the finely carved Norman doorway built into the entrance of the White Hart Hotel.

Our next day's run was short, covering only forty-two miles between Launceston and Exeter. For about half the distance the road runs along the edge of Dartmoor, the greatest of English moorlands. A motor trip of two or three days through the moor itself would be time well spent, for it abounds in romantic scenery. The road which we followed is a good one, though broken into numerous steep hills, but a part of the way we might as well have been traveling through a tunnel so far as seeing the country was concerned. A large proportion of the fences are made of earth piled up four or five feet high, and on the top of this ridge are planted the hedges, generally reaching three or four feet higher. There were times when we could catch only an occasional glimpse of the landscape, and if such fences were everywhere in England they would be a serious deterrent upon motoring. Fortunately, they prevail in a comparatively small section, for we did not find them outside of Cornwall and Devon. This experience served to impress on us how much we lost when the English landscapes were hidden—that the vistas which flitted past us as we hurried along were among the pleasantest features of our journey. It was little short of distressing to have mud fences shut from view some of the most fascinating country through which we passed.

The greatest part of the day we spent in Exeter. The Rougemont Hotel, where we stopped for the night, is spacious and comfortable, and a series of stained-glass windows at the head of the great staircase tells the story of Richard Ill's connection with Exeter; how, according to Shakespeare's play, the Rougemont of Exeter recalled to the king's superstitious mind an ancient prophecy of his defeat at the hands of Richmond, later Henry VII.

Leaving Exeter early, we planned to reach Bath in the evening—only eighty-one miles over an almost perfect road—not a very long run so far as actual distance is concerned, but entirely too long considering the places of unusual interest that lie along the way. We passed through the little town of Wellington, noted chiefly for giving his title to the Iron Duke, and it commemorates its great namesake by a lofty column reared on one of the adjacent hills.

No town in Britain has an ecclesiastical history more important than Glastonbury, whose tradition stretches back to the very beginning of Christianity in the Island. Legend has it that St. Joseph of Arimathea, who begged the body of Christ and buried it, came here in the year 63 and was the founder of the abbey. He brought with him, tradition says, the Holy Grail; and a thorn-tree staff which he planted in the abbey grounds became a splendid tree, revered for many centuries as the Holy Thorn. The original tree has vanished, though there is a circumstantial story that it was standing in the time of Cromwell and that a Puritan who undertook to cut it down as savoring of idolatry had an eye put out by a flying chip and was dangerously wounded by his axe-head flying off and striking him. With its awe-inspiring traditions—for which, fortunately, proof was not required—it is not strange that Glastonbury for many centuries was the greatest and most powerful ecclesiastical establishment in the Kingdom. The buildings at one time covered sixty acres, and many hundreds of monks and dignitaries exerted influence on temporal as well as ecclesiastical affairs. It is rather significant that it passed through the Norman Conquest unscathed; not even the greedy conquerors dared invade the sanctity of Glastonbury Abbey. The revenue at that time is said to have been about fifty thousand pounds yearly and the value of a pound then would equal twenty-five to fifty of our American dollars. However much the Normans respected the place, its sanctity had no terrors for the rapacious Henry VIII. The rich revenues appealed too strongly and he made a clean sweep, hanging the mitered abbot and two of his monks on the top of Tor Hill. The Abbey is the traditional burial-place of King Arthur and Queen Guinevere, and four of the Saxon kings sleep in unmarked graves within its precincts. Considering its once vast extent, the remaining ruins are scanty, although enough is left to show how imposing and elaborate it must have been in its palmy days. And there are few places in the Kingdom where one is so impressed with the spirit of the ancient order of things as when surrounded by the crumbling walls of Glastonbury Abbey.

ST. JOSEPH'S CHAPEL, GLASTONBURY ABBEY.

At Wells is the cathedral that gives the town an excuse for existence. Although one of the smallest of these great English churches, it is in many respects one of the most symmetrical and beautiful. Its glory is centered chiefly in its west front, with deep buttresses and many sculptured images of kings and saints. We had only an unsatisfactory glimpse of the interior, as services happened to be in progress. The town of Wells is a mere adjunct to the cathedral. It has no history of its own; no great family has ever lived there; and it can claim no glory as the birthplace of distinguished sons. Still it has a distinct charm as a quiet little Somersetshire town which has preserved its antiquity and fascination. Its name is taken from the natural wells still found in the garden of the Bishop's palace.

Bath, though it has the most remarkable Roman relics in the Kingdom, is largely modern. It is now a city of fifty thousand and dates its rise from the patronage of royalty a century and a half ago. It is one of the towns that a motorist could scarcely miss if he wished—so many fine roads lead into it—and I shall not attempt especial comment on a place so well known. Yet, as in our case, it may be a revelation to many who know of it in a general way but have no adequate idea of the real extent of the Roman baths. These date from 50 to 100 A.D. and indicate a degree of civilization which shows that the Roman inhabitants in Britain must have been industrious, intelligent and cleanly.

Excavations have been conducted with great difficulty, since the Roman remains lie directly under an important part of the city covered with valuable buildings. Nearly all of the baths in the vicinity of the springs have been uncovered and found in a surprising state of perfection. In many places the tiling with its mosaic is intact, and parts of the system of piping laid to conduct the water still may be traced. Over the springs has been erected the modern pump-house and many of the Roman baths have been restored to nearly their original state. In the pump-house is a museum with hundreds of relics discovered in course of excavation—sculpture, pottery, jewelry, coin and many other articles that indicate a high degree of civilization. Outside of the Roman remains the most notable thing in Bath is its abbey church, which, in impressive architecture and size, will compare favorably with many of the cathedrals. In fact, it originally was a cathedral, but in an early day the bishopric was transferred to Wells. There is no ruined fortress or castle in Bath, with its regulation lot of legends. Possibly in an effort to remedy the defect, there has been erected on one of the hills that overlook the town a structure which goes by the epithet of the Sham Castle.

On leaving Bath, we followed the fine London road as far as Chippenham, a prosperous agricultural town celebrated for its wool market. To the north of this is Malmesbury, with an abbey church whose history goes back to the Ninth Century. A portion of the nave is still used for services and is remarkable for its massive pillars and Norman doorway, the great arch of which has perhaps a hundred rude carvings illustrating scenes from scripture history. The strong walls of the church caused it to be used at times as a fortress, and it underwent sieges in the different wars that raged over the Kingdom. The verger pointed out to us deep indentations made by Cromwell's cannon and told us that one of the abbey's vicissitudes was its use for some years as a cloth manufacturing establishment.

IN GLOUCESTERSHIRE.

From Water Color by A. Waters.

From Malmesbury we followed the road through Cirencester to Cheltenham, one of the most modern-looking cities which we saw in England. Like Bath, it is famous for its springs, and a large share of its population is made up of retired officers of the army and navy. The main streets are very wide, nearly straight, and bordered in many places with fine trees. However, its beginning dates from only about 1700, and therefore it has little claim on the tourist whose heart is set upon ancient and historic things.

Of much greater interest is its neighbor, Gloucester, about twelve miles away. The two cities are almost of the same size, each having about fifty thousand people. Gloucester can boast of one of the most beautiful of the cathedrals, whether considered from its imposing Gothic exterior or its interior, rich with carvings and lighted by unusually fine stained-glass windows, one of which is declared to be the largest in the world. The cathedral was begun in 1088, but the main tower was not completed until nearly five hundred years later, which gives some idea of the time covered in the construction of many of these great churches. Gloucester boasts of great antiquity, for it is known that the Britons had a fortified town here which they defended against the Roman attacks; and after having become possessed of it, the Romans greatly strengthened it as a defense against incursions from the Welsh tribes. Before the Norman Conquest, it was of such importance that Edward the Confessor held his court in the town for some time. Being in the west country, it naturally was a storm-center in the parliamentary struggle, during which time a great deal of the city was destroyed. But there are many of the old portions still remaining and it has numbers of beautiful half-timbered buildings. One of these was the home of Robert Raikes, known to the world as the founder of the Sunday School. Gloucester is worthy of a longer stay than we were able to make, and in arranging an itinerary one should not fail to provide for a full day in the town.

From Gloucester to Ross runs an excellent highway, though rather devoid of interest. It was thronged with motorists who generally dashed along in sublime disregard of the speed limits. We passed several who were occupied with "road-side troubles" and we were in for an hour or so ourselves, due to a refractory "vibrator." The Welsh farmers who passed joked us good-naturedly and one said he would stick to his horse until he had money to buy a motor—then, he added, he wouldn't buy it, but would live on the income of the money. We told him that he was a man after Solomon's own heart. Suddenly the evil spirit left the car and she sprang away over the beautiful road in mad haste that soon landed us in Ross.

Ross is a pretty village, situated on a green hillside overlooking the Wye, and the tall, graceful spire of its church dominates all views of the town. Although it was growing quite late, we did not stop here, but directed our way to Monmouth, twelve miles farther on, which we reached just as the long twilight was turning into night.

DISTANT VIEW OF ROSS, SOUTH WELSH BORDER.

VIII

THROUGH BEAUTIFUL WALES

Of no part of our tour does a pleasanter memory linger than of the five or six hundred miles on the highways of Wales. The weather was glorious and no section of Britain surpassed the Welsh landscapes in beauty. A succession of green hills, in places impressive enough to be styled mountains, sloping away into wooded valleys, with here and there a quaint village, a ruined castle or abbey, or an imposing country mansion breaking on the view—all combined to make our journey through Wales one of our most pleasing experiences. Historic spots are not far apart, especially on the border, where for centuries these brave people fought English invaders—and with wonderful success, considering the greatly superior number of the aggressors. I have already written of Ludlow and Shrewsbury on the north, but scarcely less attractive—and quite as important in early days—are the fine old towns of Hereford and Monmouth on the southern border.

We were everywhere favorably impressed with the Welsh people as being thrifty and intelligent. The roadside drinking-houses were not so numerous as in England, for the Welsh are evidently more temperate in this regard than their neighbors. My observation in this particular is borne out by an English writer well qualified to judge. He says: "There is, of a truth, very little drinking now in rural Wales. The farming classes appear to be extremely sober. Even the village parliament, which in England discusses the nation's affairs in the village public house, has no serious parallel in Wales, for the detached cottage-renting laborer, who is the mainstay of such gatherings, scarcely exists, and the farmer has other interests to keep him at home." Evidently the Welsh farmer does attend to his business in an industrious manner, for he generally has a substantial and prosperous appearance. People with whom we engaged in conversation were always courteous and obliging and almost everything conspired to heighten our good opinion of the Welsh. The fusion with England is nearly complete and the Welsh language is comparatively little used except by the older people. King Edward has no more loyal subjects than the Welshmen, but apparently they do not greatly incline towards admitting his claims as their spiritual head. The Church of England in Wales is greatly inferior in numbers and influence to the various nonconformist branches. This is especially true of the more rural sections.

We found Monmouth an unusually interesting town on account of its antiquity and the numerous historic events which transpired within its walls. At the King's Head Hotel, which of course afforded shelter to Charles I when he was "touring" Britain, we were able with difficulty to find accommodation, so crowded was the house with an incursion of English trippers. Monmouth's chief glory and distinction is that it was the birthplace of King Henry V, Shakespeare's Prince Hal, whom William Watson describes as

"The roystering prince that afterward
Belied his madcap youth and proved
A greatly simple warrior lord
Such as our warrior fathers loved."

The scanty ruins of the castle where the prince was born still overlook the town. Thus King Henry became the patron of Monmouth, and in front of the town hall has been erected an inartistic effigy of a knight in full armour, with the inscription, "Henry V, born at Monmouth, August 9, 1387." The old bridge over the river Monnow is unique, with an odd, castellated gateway at one end, probably intended not so much for defense as for collecting tolls.

After dark we wandered about the streets until the church-tower chimes warned us of the lateness of the hour. And even these church bells have their history. When King Henry sailed from a seaport in France on one occasion the inhabitants rang the bells for joy, which so incensed the monarch that he ordered the bells removed and presented them to his native town. We saw too little of Monmouth, for the next morning we were away early, taking the fine road that leads directly south to Tintern and Chepstow.

The abbey-builders chose their locations with unerring judgment, always in a beautiful valley near a river or lake, surrounded by fertile fields and charming scenery. Of the score of ruined abbeys which we visited there was not one that did not fulfill this description, and none of them to a greater extent—possibly excepting Fountain's—than Tintern. In the words of an enthusiastic admirer, "Tintern is supremely wonderful for its situation among its scores of rivals. It lies on the very brink of the River Wye, in a hollow of the hills of Monmouth, sheltered from harsh winds, warmed by the breezes of the Channel—a very nook in an earthly Eden. Somehow the winter seems to fall more lightly here, the spring to come earlier, the foliage to take on a deeper green, the grass a greater thickness, and the flowers a more multitudinous variety." Certainly the magnificent church—almost entire except for its fallen roof—standing in the pleasant valley surrounded by forest-clad hills on every side, well merits such enthusiastic language. It is well that this fine ruin is now in the possession of the Crown, for it insures that decay will be arrested and its beauties preserved as an inspiration to art and architecture of later times.

From Tintern to Chepstow we followed an unsurpassed mountain road. For three miles our car gradually climbed to the highest point, winding along the hillside, from which the valley of the Severn, with its broad river, spread out beneath us in all the freshness of June verdure; while on the other hand, for hundreds of feet sheer above us, sloped the hill, with its rich curtain of forest trees, the lighter green of the summer foliage dashed with the somber gloom of the yew. Just at the summit we passed the Wyndcliffe, towering five hundred feet above us, from which one may behold one of the most famous prospects in the Island. Then our car started down a three-mile coast over a smooth and uniform grade until we landed at the brow of the steep hill which drops sharply into Chepstow.

RUINS OF RAGLAN CASTLE, SOUTH WALES.

A rude, gloomy fortress Chepstow Castle must have been in its day of might, and time has done little to soften its grim and forbidding aspect. Situated on a high cliff which drops abruptly to the river, it must have been well-nigh invincible in days ere castle walls crumbled away before cannon-shot. It is of great extent, the walls enclosing an area of about four acres, divided into four separate courts. The best-preserved portion is the keep, or tower, in which the caretaker makes his home; but the fine chapel and banqueting hall were complete enough to give a good idea of their old-time state. We were able to follow a pathway around the top of the broad wall, from which was afforded a widely extended view over the mouth of the Severn towards the sea. "This is Martin's Tower," said our guide, "for in the dungeon beneath it the regicide, Henry Martin, spent the last twenty years of his life and died." The man spoke the word "regicide" as though he felt the stigma that it carries with it everywhere in England, even though applied to the judge who condemned to death Charles Stuart, a man who well deserved to die. And when Britain punished the regicides and restored to power the perfidious race of the Stuarts, she was again putting upon herself the yoke of misgovernment and storing up another day of wrath and bloodshed.

From Chepstow it is only a short journey to Raglan, whose ruined castle impressed us in many ways as the most beautiful we saw in Britain. It was far different from the rude fortress at Chepstow. In its best days it combined a military stronghold with the conveniences and artistic effects of a palace. It is fortunately one of the best-preserved of the castellated ruins in the Kingdom. Impressive indeed were the two square towers flanking its great entrance, yet their stern aspect was softened by the heavy masses of ivy that covered them almost to the top. The walls, though roofless, were still standing, so that one could gain a good idea of the original plan of the castle. The fire places, with elaborate mantels still in place, the bits of fine carvings that clung to the walls here and there, the grand staircase, a portion of which still remains, all combined to show that this castle had been planned as a superb residence as well as a fortress. From the Gwent tower there was an unobstructed view stretching away in every direction toward the horizon. The day was perfect, without even a haze to obscure the distance, and save from Ludlow Castle, I saw nothing to equal the prospect which lay beneath me when standing on Raglan Tower.

Raglan's active history ended with its surrender August 15, 1646, to the Parliamentary army under General Fairfax, after a severe siege of more than two months. It was the last fortress in England to hold out for the lost cause of King Charles, and a brave record did its gallant defenders make against an overwhelmingly superior force. The Marquis of Worcester, though eighty-five years of age, held the castle against the Cromwellians until starvation forced him to surrender. The old nobleman was granted honorable terms by his captors, but Parliament did not keep faith, and he died a year later in the Tower of London. On being told a few days before his death that his body would be buried in Windsor Chapel, he cheerfully remarked: "Why, God bless us all, then I shall have a better castle when I am dead than they took from me when I was alive."

After the surrender the castle was dismantled by the soldiers, and the farmers in the vicinity emulated the Parliamentary destroyers in looting the fine edifice. Seventeen of the stone staircases were taken away during the interval and the great hall and chapel were seriously injured. Enough of the massive walls is left to convey a vivid idea of the olden grandeur of the castle. The motto of the time-worn arms inscribed over the entrance speaks eloquently of the past, expressing in Latin the sentiment, "I scorn to change or fear."

A quiet, unpretentious old border town is Hereford, pleasantly located on the banks of the always beautiful Wye. The square tower of the cathedral is the most conspicuous object when the town first comes into view. Though dating in part from the Eleventh Century, work on the cathedral occupied the centuries until 1530, when it was practically completed as it now stands. The vandal Wyatt, who dealt so hardly with Salisbury, had the restoration of the cathedral in hand early in the Eighteenth Century. He destroyed many of its most artistic features, but recently his work was undone and a second restoration was completed in about 1863. The structure as it now stands is mainly Norman in style, built of light-brown stone, and remarkably beautiful and imposing.

Hereford Castle has entirely vanished, though a contemporary writer describes it as "one of the fairest, largest, and strongest castles in England." The site which it occupied is now a public garden, diversified with shrubbery and flowers. An ornamental lake indicates where once was the moat, but the outlines of the walls are shown only by grass-covered ridges. Its history was no doubt as stirring as that of others of the border castles, which more fortunately escaped annihilation.

Despite its present atmosphere of peace and quietude, Hereford saw strenuous times in the fierce warfare which raged between the English and Welsh, though few relics of those days remain. The streets are unusually wide and with few exceptions the buildings are modern. Surrounding the town is a stretch of green, level meadow, upon which graze herds of the red and white cattle whose fame is wider than that of their native shire. No doubt there are many familiar with the sleek Herefords who have no idea from whence they take their name.

Our hotel, the Green Dragon, had recently been re-furnished and brightened throughout, and its excellent service was much better than we often found in towns the size of Hereford. Its well planned motor garage, just completed, showed that its proprietors recognized the growing importance of this method of touring.

Our run from Hereford up the Wye Valley to the sea, we agreed was one of our red-letter days. We passed through greatly varied scenery from the fertile, level country around Hereford to the rough, broken hills near the river's source, but the view was always picturesque in the highest degree. The road runs along the edge of the hills, and the glorious valley with its brawling river spread out before us almost the entire day. At times we ran through forests, which cover the immense parks surrounding the country estates along the river. We saw many fine English country-seats, ranging from old, castellated structures to apparently modern mansions. There are also a number of ruins along the valley, each with its romantic legends. At Hay, on the hill overlooking the town, is the castle, partly in ruins and partly in such state of repair as to be the summer home of the family that owns it. A little farther, upon a knoll directly overhanging the river, are crumbling piles of stone where once stood Clifford Castle, the home of Fair Rosamond, whose melancholy story Tennyson has woven into one of his dramas.

As we advanced farther up the valley, the country grew wilder and more broken and for many miles we ran through the towering hills that pass for mountains in Wales. These were covered with bright-green verdure to their very tops, and the flocks of sheep grazing everywhere lent an additional charm to the picture. At the foot of the hills the road follows the valleys with gentle curves and easy grades. The Wye dwindles to the merest brook, and some miles before we reached the coast, we passed the head waters of the river and followed a brook flowing in an opposite direction.

The road over which we had traveled is not favorable for fast time. Though comparatively level and with splendid surface, it abounds in sharp curves and in many places runs along high embankments. The Motor Union has recommended that eighteen miles per hour be not exceeded on this road. The distance from Hereford to Aberyswith is only ninety miles, yet we occupied the greater part of the day in the trip, and had time permitted, we would gladly have broken the journey at one of the quaint towns along the way. At many points of vantage we stopped to contemplate the beauty of the scene — one would have to be a speed maniac indeed to "scorch" over the Wye Valley road.

Aberyswith is a seaside resort, somewhat similar to Penzance. It is situated on the harbor at the foot of a high bluff, and its principal feature is the long row of hotels fronting on the ocean. Though mostly modern, it is by no means without history, as evidenced by its ruined castle overlooking the sea and vouching for the antiquity of the town.

We left Aberyswith next morning with considerable apprehensions. Our books and maps showed that we would encounter by odds the worst roads of our entire tour. A grade of one in five along the edge of an almost precipitous hill was not an alluring prospect, for we were little inclined toward hill-climbing demonstrations. Shortly after leaving the town we were involved in poorly kept country byways without sign-boards and slippery with heavy rains of the night before. After meandering among the hills and inquiring of the natives for towns the names of which they could not understand when we asked and we could not understand when they answered, we came to Dinas Mowddwy, where there was little else than a handsome hotel. This reminded us that in our wanderings the hour for luncheon had passed. We stopped at the hotel, but found difficulty in locating anybody to minister to our wants; and so deliberate were the movements of the party who finally admitted responsibility that an hour was consumed in obtaining a very unpretentious repast.

The hotelkeeper held out a discouraging prospect in regard to the hills ahead of us. He said that the majority of the motorists who attempted them were stalled and that there had been some serious accidents. We went on our way with considerable uneasiness, as our car had not been working well, and later on trouble was discovered in a broken valve-spring. However, we started over the mountain, which showed on our road-book to be not less than three miles in length. There were many dangerous turns of the road, which ran alongside an almost precipitous incline, where there was every opportunity for the car to roll a mile or more before coming to a standstill if it once should get over the edge. We crawled up the hill until within about fifty yards from the top, and right at this point there was a sharp turn on an exceedingly stiff grade. After several trials at great risk of losing control of the car, I concluded that discretion was (sometimes) the better part of valor, and with great difficulty turned around and gave it up.

We made a detour by way of Welshpool and Oswestry, where we came into the London and Holyhead road, bringing up for the night at Llangollen. We found it necessary to travel about sixty miles to get to the point which we would have reached in one-fourth the distance had we succeeded in climbing the hill. It proved no hardship, as we saw some of the most beautiful country in Wales and traveled over a level road which enabled us to make very good time with the partly crippled car.

Although Llangollen is a delightful town, my recollections of it are anything but pleasant. Through our failure to receive a small repair which I ordered from London, we were delayed at this place for two days, and as it usually chances in such cases, at one of the worst hotels whose hospitality we endured during our trip. It had at one time been quite pretentious, but had degenerated into a rambling, dirty, old inn, principally a headquarters for fishing parties and local "trippers." And yet at this dilapidated old inn there were a number of guests who made great pretensions at style. Women "dressed for dinner" in low-necked gowns with long trains; and the men attired themselves in dress-suits of various degrees of antiquity.

While we were marooned here we visited Vale Crucis Abbey, about a mile distant. The custodian was absent, or in any event could not be aroused by vigorously ringing the cowbell suspended above the gate, and we had to content ourselves with a very unsatisfactory view of the ruin over the stone wall that enclosed it. The environments of Llangollen are charming in a high degree. The flower-bordered lanes lead past cottages and farm houses surrounded by low stone walls and half hidden by brilliantly colored creepers. Bits of woodland are interspersed with bright green sheep pastures and high, almost mountainous, bluffs overhang the valley. On the very summit of one of these is perched a ruined castle, whose inaccessible position discouraged nearer acquaintance.

The country around Llangollen was beautiful, but the memory of the hotel leaves a blight over all. We were happy indeed when our motor started off again with the steady, powerful hum that so delights the soul of the driver, and it seemed fairly to tremble with impatience to make up for its enforced inaction. Though it was eight o'clock in the evening, it was anything to get away from Llangollen, and we left with a view of stopping for the night at Bettws-y-Coed, about thirty miles away.

With our motor car racing like mad over the fine highway—there was no danger of police traps at that hour—we did not stop to inquire about the dog that went under the wheels in the first village we passed. However, the night set in suddenly and a rain began to fall heavily before we had gone half the distance we proposed. We had experienced trouble enough in finding the roads in Wales during the daytime, and the prospect of doing this by night and in a heavy rain was not at all encouraging, and we perforce had to put up at the first place that offered itself. A proposition to stop at one of the so-called inns along the road was received with alarm by the good woman who attended the bar. She could not possibly care for us and she was loud in her praises of the Saracen's Head at Cerrig-y-Druidion, only a little farther on, which she represented as a particular haven for motorists.

The appearance of our car with its rapidly vibrating engine and glaring headlights before the Saracen's Head created considerable commotion among the large family of the host and the numerous guests, who, like Tam-O'-Shanter, were snug and cozy by their inglenook while the storm was raging outside. However, the proprietor was equal to the occasion and told me that he had just come from Liverpool to take charge of the inn and that he hoped to have the patronage of motorists. With commendable enterprise he had fitted up a portion of his barn and had labeled it "Motor Garage" in huge letters. The stable man was also excited over the occasion, and I am sure that our car was the first to occupy the newly created garage, which had no doubt been cut off from the cow-stable at a very recent date.

The shelter of the Saracen's Head was timely and grateful none the less, and no one could have been kindlier or more attentive than our hostess. We had a nicely served lunch in the hotel parlor, which was just across the hallway from the lounging room, where the villagers assembled to indulge in such moderate drinking as Welshmen are addicted to. The public room was a fine old apartment with open-beamed ceiling—not the sham with which we decorate our modern houses, but real open beams that supported the floor—and one end of the room was occupied by a great open fireplace with old-time spits and swinging cranes. Overhead was hung a supply of hams and bacon and on iron hooks above the door were suspended several dressed fowls, on the theory that these improve with age. We were given a small but clean and neat apartment, from which I suspicion the younger members of the landlord's family had been unceremoniously ousted to make room for us. The distressing feature was the abominable beds, but as these prevailed in most of the country hotels at which we stopped we shall not lay this up too strongly against the Saracen's Head. I noticed that on one of the window-panes someone had scribbled with a diamond, "Sept. 4, 1726," which would seem to indicate that the original window was there at that time. The house itself must have been considerably older. If rates had been the sole inducement, we should undoubtedly have become permanent boarders at the Saracen's Head, for I think that the bill for our party was seven shillings for supper, room and breakfast.

We left Cerrig-y-Druidion next morning in a gray, driving rain, with drifting fogs that almost hid the road at times. A few miles brought us to the Conway River, the road closely following the stream through the picturesque scenery on its banks. It was swollen by heavy rains and the usually insignificant river was a wild torrent, dashing in rapids and waterfalls over its rocky bed. The clouds soon broke away and for the remainder of the day the weather was as fine as could possibly be wished for.

Bettws-y-Coed is the most famous of mountain towns in Wales, and its situation is indeed romantic. It is generally reputed to be the chief Welsh honeymoon resort and a paradise for fishermen, but it has little to detain the tourist interested in historic Britain. We evidently should have fared much differently at its splendid hotel from what we did at Cerrig-y-Druidion, but we were never sorry for our enforced sojourn at the Saracen's Head.

The road from Bettws-y-Coed to Carnarvon is a good one, but steep in places, and it passes through some of the finest mountain scenery in Wales. It leads through the Pass of Llanberis and past Snowdon, the king of the Welsh mountains—though tame indeed to one who has seen the Rockies. Snowdon, the highest in the Kingdom, rises not so much as four thousand feet above the sea level.

Carnarvon Castle is conceded from many points of view to be the finest ruin in the Kingdom. It does not occupy an eminence, as did so many castles whose position contributed much to their defense, but it depended more on its lofty watch-towers and the stupendous strength of its outer walls. These are built of solid granite with a thickness of ten feet or more in vital places, and it is doubtful if even the old-time artillery would have made much impression upon them. Its massive construction no doubt accounts for the wonderful preservation of the outer walls, which are almost entire, and Carnarvon Castle, as viewed from the outside, probably appears very much the same as it did when the builders completed the work about 1300. It was built by King Edward I as a royal residence from which to direct his operations against the Welsh, which finally resulted in the conquest of that people by the English invaders. In a little dungeonlike room, tradition declares that Edward II, first Prince of Wales, was born. This is vigorously insisted upon in the local guide-book as an actual historic fact, although it is quite as vigorously disputed by numerous antiquarians, uninfluenced by Carnarvon's interests. The castle is now the property of the town and is well looked after.

Leaving Carnarvon, our next objective was Conway, whose castle is hardly less famous and even more picturesque than that of its neighbor, though in more ruinous condition. The road we followed closely skirts the coast for a great part of the distance, running at times on the verge of the ocean. In places it reminds one of the Axenstrasse of Lake Lucerne, being cut in the side of the cliffs overhanging the sea, with here and there great masses of rock projecting over it; and passes occasionally through a tunnel cut in the stone. A few miles north of Carnarvon we passed through Bangor, one of the most prosperous-looking towns in North Wales and the seat of one of the few Welsh cathedrals—a long, low, though not unpleasing, building. The site of this cathedral had been continuously occupied by a church since the Sixth Century, although the present structure dates from the Thirteenth.

An hour's run after leaving Bangor brought us in sight of the towers of Conway Castle. Nowhere in Britain does the spirit of mediaevalism linger as it does in the ancient town of Conway. It is still surrounded by its old wall with twenty-one watch-towers and the three gateways originally leading into the town have been recently restored. The castle stands on the verge of a precipitous rock and its outer walls are continuous with those of the town. It is a perfect specimen of a Thirteenth Century military fortress, with walls of enormous thickness, flanked by eight huge, circular towers. It was built by Edward I in 1284. Several times it was besieged by the Welsh and on one occasion came near falling into their hands while the king himself was in the castle. It was besieged during the Parliamentary wars, but for some unaccountable reason it was not destroyed or seriously damaged when captured. Its present dilapidated state is due to the action of its owner, Lord Conway, shortly after, in dismantling it to sell the lead and timber of the building, and it was permitted to fall into gradual decay. The castle, with its eight towers and bridge, which matches it in general style and which was built about fifty years ago, is one of the best known objects in the whole Kingdom. It has been made familiar to everybody through innumerable photographs and pictures.

When we drew our car up in front of the castle it was in gala attire and was the scene of activity which we were at a loss to account for. We soon learned that the Wesleyans, or Welsh Methodists, were holding a festival in the castle, and the shilling we paid for admission included a nicely served lunch, of which the Welsh strawberries were the principal feature. The occasion was enlivened by music from the local band and songs by young girls in the old Welsh costume. This led us to ask if the Welsh language were in common use among the people. We were told that while the older people can speak it, it does not find much favor among the younger generation, some of whom are almost ashamed to admit knowledge of the old tongue. English was spoken everywhere among the people at the gathering, and the only Welsh heard was in some of the songs by the girls. We wandered about the ruin and ascended the towers, which afford a fine view of the town and river. There seems to have been little done in the way of restoration, or repair, but so massive are the walls that they have splendidly stood the ravages of time.

On leaving Conway we crossed the suspension bridge, paying a goodly toll for the privilege. It was already growing late when we left the town, but the fine level road and the unusually willing spirit evinced by our motor enabled us to cover the fifty miles to Chester before night set in.

IX

CHESTER TO "THE HIELANDS"

Chester stands a return visit well, and so does the spacious and hospitable Grosvenor Hotel. It was nearly dark when we reached the city and the hotel was crowded, the season now being at its height. We had neglected to wire for reservation, but our former stop at the hotel was not forgotten and this stood us in good stead in securing accommodations. So comfortably were we established that we did not take the car out of the garage the next day but spent our time in leisurely re-visiting some of the places that had pleased us most.

The next day we were early away for the north. I think that no other stretch of road of equal length was more positively unattractive than that we followed from Chester to Penrith. Even the road-book, whose "objects of interest" were in some cases doubtful, to say the least, could name only the battlefield of 1648 near Preston and one or two minor "objects" in a distance of one hundred miles. I recalled the comment of the Touring Secretary of the Motor Union as he rapidly drew his pencil through this road as shown on the map: "Bad road, rough pavement, houses for thirty miles at a stretch right on each side of the street, crowds of children everywhere—but you cannot get away from it very well." All of which we verified by personal experience.

At starting it seemed easy to reach Carlisle for the night, but progress was slow and we met an unexpected delay at Warrington, twenty miles north of Chester. A policeman courteously notified us that the main street of the city would be closed three hours for a Sunday School parade. We had arrived five minutes too late to get across the bridge and out of the way. We expressed our disgust at the situation and the officer made the conciliatory suggestion that we might be able to go on anyway. He doubted if the city had any authority to close the main street, one of the King's highways, on account of such a procession. We hardly considered our rights so seriously infringed as to demand such a remedy, and we turned into the stable-yard of a nearby hotel to wait until the streets were clear. In the meantime we joined the crowd that watched the parade. The main procession, of five or six thousand children, was made up of Sunday Schools of the Protestant churches—the Church of England and the "Non-Conformists." The Catholics, whose relations in England with Protestants are strained to a much greater extent than in the United States, did not join, but formed a smaller procession in one of the side streets. The parade was brilliant with flags and with huge banners bearing portraits of the King and Queen, though some bore the names and emblems of the different schools. One small fellow proudly flourished the Stars and Stripes, which was the only foreign flag among the thousands in the procession. In this connection I might remark that one sees the American flag over here far oftener than he would traveling in America. We found nothing but the kindest and most cordial feeling toward Americans everywhere; and the very fact that we were Americans secured us special privileges in not a few cases.

After the procession had crossed the bridge, a policeman informed us that we could proceed. We gained considerable time by making a detour through side streets—not an altogether easy performance—and after much inquiry regained the main road leading out of the city. Warrington is a city of more than one hundred and twenty thousand inhabitants, a manufacturing place with nothing to detain the tourist. On the main street near the river is a fine bronze statue of Oliver Cromwell, one of four that I saw erected to the memory of the Protector in England. Our route from Warrington led through Wigan and Preston, manufacturing cities of nearly one hundred thousand each, and the suburbs of the three are almost continuous. Tram cars were numerous and children played everywhere with utter unconcern for the vehicles which crowded the streets.

When we came to Lancaster we were glad to stop, although our day's journey had covered only sixty miles. We knew very little of Lancaster and resorted to the guide-books for something of its antecedents, only to learn the discouraging fact that here, as everywhere, the Romans had been ahead of us. The town has a history reaching back to the Roman occupation, but its landmarks have been largely obliterated in the manufacturing center which it has become. Charles Dickens was a guest at Lancaster, and in recording his impressions he declared it "a pleasant place, dropped in the midst of a charming landscape; a place with a fine, ancient fragment of a castle; a place of lovely walks and possessing many staid old houses, richly fitted with Honduras mahogany," and followed with other reflections not so complimentary concerning the industrial slavery which prevailed in the city a generation or two ago. The "fine, ancient fragment of a castle" has been built into the modern structure which now serves as the seat of the county court. The square tower of the Norman keep is included in the building. This in general style and architecture conforms to the old castle, which, excepting the fragment mentioned by Dickens, has long since vanished. Near at hand is St. Mary's Church, rivaling in size and dignity many of the cathedrals, and its massive, buttressed walls and tall, graceful spire do justice to its magnificent site. From the eminence occupied by the church the Irish Sea is plainly visible, and in the distance the almost tropical Isle of Man rises abruptly out of the blue waters. The monotony of our previous day's travel was forgotten in lively anticipation as we proceeded at what seemed a snail's pace over the fine road leading from Penrith to Carlisle. We had been warned at Penrith, not against the bold highwaymen, the border moss-troopers or the ranting Highlandmen of song and story, but against a plain, Twentieth Century police trap which was being worked very successfully along this road. Such was our approach in these degenerate days to "Merrie Carlile," which figured so largely in the endless border warfare between the Scotch and English. But why the town should have been famed as "Merrie Carlile" would be hard to say, unless more than a thousand years of turmoil, bloodshed and almost ceaseless warfare through which it passed earned it the cheerful appellation. The trouble between the English and the Welsh ended early, but it has been only a century and a half ago since the closing scene of the long and bitter conflict between the north and south was enacted at Carlisle. Its grim old castle was the scene of the imprisonment and execution of the last devoted followers of Prince Charlie, and according to Scott's Waverly the dashing but sadly deluded young chieftain, Fergus McIvor, was one of those who suffered a shameful death. In this connection one remembers that Scott's marriage to Miss Carpentier took place in Carlisle, an event that would naturally accentuate our interest in the fine old border city. As we had previously visited Carlisle, our stay was a short one, but its remarkable history, its connection with the stories of Walter Scott, its atmosphere of romance and legend and the numerous points of interest within easy reach—all combine to make it a center where one might spend several days. The Romans had been here also, and they, too, had struggled with the wild tribes on the north, and from that time down to the execution of the last adherents of the Stuarts in 1759 the town was hardly at any time in a state of quietude. As described by an observant writer, "every man became a soldier and every house that was not a mere peasant's hut was a fortress." A local poet of the Seventeenth Century summed it up in a terse if not elegant couplet as his unqualified opinion

"That whoso then in the border did dwell
Lived little happier than those in hell."

But Carlisle is peaceful and quiet enough at the present time, a place of considerable size and with a thriving commerce. Its castle, a plain and unimpressive structure, still almost intact, has been converted into military barracks, and its cathedral, which, according to an old chronicle, in 1634 "impressed three observant strangers as a great wild country church," has not been greatly altered in appearance since that period. It suffered severely at the hands of the Parliamentary soldiers, who tore down a portion of the nave to use the materials in strengthening the defenses of the town. But the story of Carlisle could not be told in many volumes. If the mere hint of its great interest which I have given here can induce any fellow tourist to tarry a little longer at "Merrie Carlile," it will be enough.

Leaving Carlisle, we crossed "Solway Tide" and found ourselves in the land of bluebells and heather, the "Bonnie Scotland" of Robert Burns. Shortly after crossing the river, a sign-board pointed the way to Gretna Green, that old-time haven of eloping lovers, who used to cross the Solway just as the tide began to rise, and before it subsided there was little for the paternal ancestors to do but forgive and make the best of it. But we missed the village, for it was a mile or two off the road to Dumfries, which we hoped to reach for the night. An unexpected difficulty with the car nearly put this out of the range of possibility, but by grace of the long Scotch twilight, we came into Dumfries about ten o'clock without finding it necessary to light our lamps. Our day's journey had been a tiresome one, and we counted ourselves fortunate on being directed to the Station Hotel, which was as comfortable and well managed as any we found. The average railway hotel in America is anything but an attractive proposition, but in Scotland and in England conditions are almost reversed, the station hotels under the control of the different railway companies being generally the best.

ENTRANCE TO LOCH TYNE.

From Water Color by Stewart.

We had been attracted to Dumfries chiefly because of its association with Robert Burns, who spent the last years of his life in the town or in its immediate vicinity. Our first pilgrimage was to the poet's tomb, in St. Michael's churchyard. A splendid memorial marks the place, but a visit to the small dingy house a few yards distant, in which he died, painfully reminded us of his last years of distress and absolute want. Within easy reach of Dumfries lie many points of interest, but as our time permitted us to visit only one of these, we selected Caerlaverock Castle, the Ellangowan of Scott's "Guy Mannering," lying about ten miles to the south. In location and style of construction it is one of the most remarkable of the Scotch ruins. It stands in an almost level country near the coast and must have depended for defense on its enormously thick walls and the great double moat which surrounded it, rather than the strength of its position. The castle is built of dark-brown stone, and the walls, rising directly from the waters of the moat and covered with masses of ivy, are picturesque, though in a sad state of disrepair. Bits of artistic carving and beautiful windows showed that it was a palace as well as a fortress, though it seems strange that the builder should select such a site. In common with most British castles, it was finally destroyed by Cromwell, and the custodian showed us a pile of cannon balls which he had gathered in the vicinity. On one of the stones of the inner wall were the initials, "R.B.," and the date, "1776," which our guide assured us were cut by Robert Burns; and there are certain peculiarities about the monogram which leave little doubt that it was the work of the poet. From the battlements of the castle the old man pointed to a distant hill, where, he told us, the home of the Carlyles had been for many years and where Thomas Carlyle, who was born at Ecclefechan, lies buried. Within a few miles of Dumfries is Ellisland Farm, where Robert Burns was a tenant for several years, and many of his most famous poems were written during that period. And besides, there were old abbeys and castles galore within easy reach; and glad indeed we should have been had we been able to make the Station Hotel our headquarters for a week and devote our time to exploring. But we were already behind schedule and the afternoon found us on the road to Ayr.

A little more than half the distance from Dumfries to Ayr the road runs through the Nith Valley, with river and forest scenery so charming as to remind us of the Wye. The highway is a splendid one, with fine surface and easy grades. It passes through an historic country, and the journey would consume a long time if one should pause at every point that might well repay a visit. A mile on the way is Lincluden Abbey, in whose seclusion Burns wrote many of his poems, the most famous of which, "The Vision of Liberty," begins with a reference to the ruin:

> "As I stood by yon roofless tower
> Where wall flowers scent the dewy air,
> Where the owlet lone in her ivy bower,
> Tells to the midnight moon her care—"

Ellisland Farm is only a few miles farther on the road, never to be forgotten as the spot where "Tam-O'-Shanter" was written. The farm home was built by Burns himself during what was probably the happiest period of his life, and he wrote many verses that indicated his joyful anticipation of life at Ellisland Farm. But alas, the "best laid plans o' mice and men gang oft agley," and the personal experience of few men has more strikingly proven the truth of the now famous lines than of Robert Burns himself! Many old castles and magnificent mansions crown the heights overlooking the river, but we caught only glimpses of some of them, surrounded as they were by immense parks, closed to the public. Every one of the older places underwent many and strange vicissitudes in the long years of border warfare, and of them all, Drumlanrigh Castle, founded in 1689, is perhaps the most imposing. For ten years its builder, the first Earl of Queensbury, labored on the structure, only to pass a single night in the completed building, never to revisit it, and ending his days grieving over the fortune he had squandered on this many-towered pile of gray stone.

We may not loiter along the Nithdale road, rich as it is in traditions and relics of the past. Our progress through such a beautiful country had been slow at the best, and a circular sign-board, bearing the admonition, "Ten Miles Per Hour," posted at each of the numerous villages on the way, was another deterrent upon undue haste. The impression that lingers with us of these small Scotch villages is not a pleasant one. Rows of low, gray-stone, slate-roofed cottages straggling along a single street—generally narrow and crooked and extending for distances depending on the size of the place—made up the average village. Utterly unrelieved by the artistic touches of the English cottages and without the bright dashes of color from flowers and vines, with square, harsh lines and drab coloring everywhere, these Scotch villages seemed bleak and comfortless. Many of them we passed through on this road, among them Sandquhar, with its castle, once a strong and lordly fortress but now in a deplorable state of neglect and decay, and Mauchline, where Burns farmed and sang before he removed to Dumfries. It was like passing into another country when we entered Ayr, which, despite its age and the hoary traditions which cluster around it, is an up-to-date appearing seaport of about thirty thousand people. It is a thriving business town with an unusually good electric street-car system, fine hotels and (not to be forgotten by motorists) excellent garages and repair shops.

Ayr is one of the objective points of nearly every tourist who enters Scotland. Its associations with Burns, his birthplace, Kirk Alloway, his monument, the "Twa Brigs," the "Brig O' Doon," and the numerous other places connected with his memory in Ayr and its vicinity, need not be dwelt on here. An endless array of guide-books and other volumes will give more information than the tourist can absorb and his motor car will enable him to rapidly visit such places as he may choose. It will be of little encumbrance to him, for he may leave the car standing at the side of the street while he makes a tour of the haunts of Burns at Alloway or elsewhere.

It was a gloomy day when we left Ayr over the fine highway leading to Glasgow, but before we had gone very far it began to rain steadily. We passed through Kilmarnock, the largest city in Ayrshire. Here a splendid memorial to Burns has been erected, and connected with it is a museum of relics associated with the poet, as well as copies of various editions of his works. This reminds one that the first volume of poems by Burns was published at Kilmarnock, and in the cottage at Ayr we saw one of the three existing copies, which had been purchased for the collection at an even thousand pounds.

THE PATH BY THE LOCH.

From Photograph.

We threaded our way carefully through Glasgow, for the rain, which was coming down heavily, made the streets very slippery, and our car showed more or less tendency to the dangerous "skid." Owing to former visits to the city, we did not pause in Glasgow, though the fact is that no other large city in Britain has less to interest the tourist. It is a great commercial city, having gained in the last one hundred years three quarters of a million inhabitants. Its public buildings, churches, and other show-places—excepting the cathedral—lack the charm of antiquity. After striking the Dumbarton road, exit from the city was easy, and for a considerable distance we passed near the Clyde shipyards, the greatest in the world, where many of the largest merchant and war vessels have been constructed. Just as we entered Dumbarton, whose castle loomed high on a rocky island opposite the town, the rain ceased and the sky cleared with that changeful rapidity we noticed so often in Britain. Certainly we were fortunate in having fine weather for the remainder of the day, during which we passed perhaps as varied and picturesque scenery as we found on our journey.

For the next thirty miles the road closely followed the west shore of Loch Lomond, and for the larger part of the way we had a magnificent panorama of the lake and the numberless green islands that rose out of its silvery waters. Our view in places was cut off by the fine country estates that lay immediately on the shores of the lake, but the grounds, rich with shrubbery and bright with flowers, were hardly less pleasing than the lake itself. These prevailed at the southern portion of the lake only, and for at least twenty miles the road closely followed the shore, leading around short turns on the very edges of steep embankments or over an occasional sharp hill—conditions that made careful driving necessary. Just across the lake, which gradually grew narrower as we went north, lay the low Scotch mountains, their green outlines subdued by a soft blue haze, but forming a striking background to the ever-varying scenery of the lake and opposite shore. Near the northern end on the farther side is the entrance to the Trosachs, made famous by Scott's "Lady of the Lake." The roads to this region are closed to motors—the only instance that I remember where public highways were thus interdicted. The lake finally dwindled to a brawling mountain stream, which we followed for several miles to Crianlarich, a rude little village nestling at the foot of the rugged hills. From here we ran due west to Oban, and for twenty miles of the distance the road was the worst we saw in Scotland, being rough and covered with loose, sharp stones that were ruinous to tires. It ran through a bleak, unattractive country almost devoid of habitations and with little sign of life excepting the flocks of sheep grazing on the short grasses that covered the steep, stony hillsides. The latter half of the distance the surroundings are widely different, an excellent though winding and narrow road leading us through some of the finest scenes of the Highlands. Especially pleasing was the ten-mile jaunt along the north shore of Loch Awe, with the glimpses of Kilchurn Castle which we caught through occasional openings in the thickly clustered trees on the shore. Few ruins are more charmingly situated than Kilchurn, standing as it does on a small island rising out of the clear waters—the crumbling walls overgrown with ivy and wall-flowers. The last fifteen miles were covered in record time for us, for it was growing exceedingly chilly as the night began to fall and the Scotch July day was as fresh and sharp as an American

October.

KILCHURN CASTLE, LOCH AWE.

Oban is one of the most charming of the north of Scotland resort towns, and is becoming one of the most popular. It is situated on a little land-locked bay, generally white in summer time with the sails of pleasure vessels. Directly fronting the town, just across the harbor, are several ranges of hills fading away into the blue mists of the distance and forming, together with the varying moods of sky and water, a delightful picture. Overhanging the town from the east is the scanty ruin of Dunollie Castle, little more than a shapeless pile of stone covered over with masses of ivy. Viewed from the harbor, the town presents a striking picture, and the most remarkable feature is the great colosseum on the hill. This is known as McCaig's Tower and was built by an eccentric citizen some years ago merely to give employment to his fellow townsmen. One cannot get an adequate idea of the real magnitude of the structure without climbing the steep hill and viewing it from the inside. It is a circular tower, pierced by two rows of windows, and is not less than three hundred feet in diameter, the wall ranging in height from thirty to seventy-five feet from the ground. It lends a most striking and unusual appearance to the town, but among the natives it goes by the name of "McCaig's Folly."

From Oban as a center, numberless excursions may be made to old castles, lakes of surpassing beauty and places of ancient and curious history. None of the latter are more famous than the island of Iona, lying about thirty-five miles distant and accessible by steamer two or three days of each week in summer time. We never regretted that we abandoned the car a day for the trip to this quaint spot and its small sister island, Staffa, famed for Fingall's cave and the curious natural columns formed by volcanic action. The round trip covers a distance of about seventy-five miles and occupies eight or ten hours. Iona is a very small island, with a population of no more than fifty, but it was a place of importance in the early religious history of Scotland; and its odd little cathedral, which is now in ruins—except the nave, but recently restored—was originally built in the Eleventh Century. Weird and strange indeed is the array of memorials rudely cut from Scotch granite that mark the resting places of the chiefs of many forgotten clans, while a much higher degree of art is shown in the regular and even delicate designs traced on the numerous old crosses still standing. In olden days Iona was counted sacred ground after the landing of St. Columba in 563, and its fame even extended to Sweden and Denmark, whose kings at one time were brought here for interment. We were fortunate in having a fine day, the sky being clear and the sea perfectly smooth. We were thus enabled to make landing at both isles, a thing that is often impossible on account of the weather. This circular trip—for the return is made by the Sound of Mull—is a remarkably beautiful one, the steamer winding in and out through the straits among the islands and between shores wild and broken, though always picturesque and often impressive. Many of the hills are crowned with ruined fortresses and occasionally an imposing modern summer residence is to be seen. Competent judges declare that provided the weather is fine no more delightful short excursion by steamer can be made on the British coast than the one just described. Three miles from Oban lies Dunstafnage Castle, a royal residence of the Pictish kings, bearing the marks of extreme antiquity. It occupies a commanding position on a point of land extending far into the sea and almost surrounded by water at high tide. We visited it in the fading twilight, and a lonelier, more ghostly place it would be hard to imagine. From this old castle was taken the stone of destiny upon which the Pictish kings were crowned, but which is now the support of the coronation chair in Westminster Abbey. A place so rich in romantic legend could not be expected to escape the knowledge of the Wizard of the North and Scott made more than one visit to this solitary ruin. As a result the story of Dunstafnage has been woven into the "Legend of Montrose" as "Ardenvohr" and the description may be easily recognized by any one who visits the old castle.

Oban is modern, a place of many and excellent hotels fronting on the bay. So far, only a small per cent of its visitors are Americans, and the indifferent roads leading to the town discourage the motorist. Had we adhered to the route outlined for us by the Motor Union Secretary, we should have missed it altogether. We had made a stop in the town two years before, and yet there are few places in Britain that we would rather visit a third time than Oban.

X

THROUGH HISTORIC SCOTLAND

The north of Scotland is rapidly becoming little more than a pleasure-ground for the people of the Kingdom, and its attractions are yearly drawing a larger number of Americans. There are practically no European visitors, but that is largely true of the entire Kingdom. The people of the Continent consider Britain a chilly, unattractive land. Its historic and literary traditions, so dear to the average American, who holds a common language, do not appeal to those who think their own countries superior to any other in these particulars.

It is only a natural consequence that Scotland, outside of the three or four largest cities, is becoming, like Switzerland, a nation of hotelkeepers—and very excellent ones they are. The Scotch hotels average as good as any in the world. One finds them everywhere in the Highlands. Every lake, every ruin frequented by tourists has its hotel, many of them fine structures of native granite, substantially built and splendidly furnished.

We left Oban over the route by which we came, since no other was recommended to motorists. Our original plan to follow the Caledonian Canal to Inverness was abandoned on account of difficult roads and numerous ferries with poor and infrequent service. After waiting three hours to get an "accumulator" which had been turned over to a local repair man thirty-six hours before with instructions to have it charged and returned promptly, we finally succeeded in getting off. This delay is an example of those which we encountered again and again from failure to get prompt service, especially when we were making an effort to get away before ten or eleven in the morning.

It was no hardship to follow more leisurely than before the road past Loch Awe, whose sheet of limpid water lay like a mirror around Kilchurn Castle under the cloudless, noonday sky. A little farther on, at Dalmally, we paused at a pleasant old country hotel, where the delicious Scotch strawberries were served fresh from the garden. It was a quaint, clean, quiet place, and the landlord told us that aside from the old castles and fine scenery in the vicinity, its chief attraction to guests was trout-fishing in neighboring streams. We were two days in passing through the heart of the Highlands from Oban to Inverness over about two hundred miles of excellent road running through wild and often beautiful scenery, but there were few historic spots as compared with the coast country. The road usually followed the edge of the hills, often with a lake or mountain stream on one hand. From Crianlarich we followed the sparkling Dochart until we reached the shore of Loch Tay, about twenty miles distant. From the mountainside we had an unobstructed view of this narrow but lovely lake, lying for a distance of twenty miles between ridges of sharply rising hills. White, low-hung clouds half hid the mountains on the opposite side of the loch, giving the delightful effect of light and shadow for which the Scotch Highlands are famous and which the pictures of Watson, Graham and Farquharson have made familiar to nearly everyone.

At the northern end of the lake we caught distant glimpses of the battlemented towers of Taymouth Castle, home of the Marquis of Breadalbane, which, though modern, is one of the most imposing of the Scotch country seats. If the castle itself is imposing, what shall we say of the estate, extending as it does westward to the Sound of Mull, a distance of one hundred miles—a striking example of the inequalities of the feudal system. Just before we crossed the bridge over the Tay River near the outlet of the lake, we noticed a gray old mansion with many Gothic towers and gables, Grandtully Castle, made famous by Scott as the Tully-Veolan of Waverly. Near by is Kinniard House, where Robert Louis Stevenson wrote "Treasure Island."

A few miles farther on we came to Pitlochry, a surprisingly well built resort with excellent hotels and a mammoth "Hydropathic" that dominates the place from a high hill. The town is situated in the very center of the Highlands, surrounded by hills that supply the gray granite used in its construction; and here we broke our journey for the night.

Our way to Inverness was through a sparsely inhabited, wildly broken country, with half a dozen mean-looking villages at considerable distances from each other and an occasional hut or wayside inn between. Although it was July and quite warm for the north of Scotland, the snow still lingered on many of the low mountains, and in some places it seemed that we might reach it by a few minutes' walk. There was little along the road to remind one of the stirring times or the plaided and kilted Highlander that Scott has led us to associate with this country. We saw one old man, the keeper of a little solitary inn in the very heart of the hills, arrayed in the full glory of the old-time garb—plaid, tartan, sporran and skene-dhu, all set off by the plumed Glengarry cap—a picturesque old fellow indeed. And we met farther on the way a dirty-looking youth with his bagpipes slung over his shoulder—in dilapidated modern garb he was anything but a fit descendant of the minstrels whose fame has come down to us in song and story. Still, he was glad to play for us, and despite his general resemblance to an every-day American tramp, it was something to have heard the skirl of the bag-pipe in the Pass of Killiekrankie. And after all, the hills, the vales and the lochs were there, and everywhere on the low green mountains grazed endless flocks of sheep. They lay leisurely in the roadway or often trotted unconcernedly in front of the car, occasioning at times a speed limit even more unsatisfactory than that imposed in the more populous centers by the police traps. Incidentally we learned that the finest sheep in the world—and vast numbers of them—are produced in Great Britain. When we compare them with the class of animals raised in America it is easy to see why our wool and mutton average so greatly inferior.

IN THE SCOTTISH HIGHLANDS.

From Painting by D. Sherrin.

A clean, quiet, charming city is Inverness, "the capital of the Highlands," as the guide-books have it. It is situated on both shores of its broad, sparkling river—so shallow that the small boys with turned-up pantaloons wade across it in summer time—while an arm of the sea defines the boundary on the northeast. Though tradition has it that Macbeth built a castle on the site of the present structure, it disappeared centuries ago, and there is now little evidence of antiquity to be found in the town. The modern castle is a massive, rambling, brown-stone building less than a hundred years old, now serving as a county court. The cathedral is recent, having been completed in the last quarter of a century. It is an imposing church of red stone, the great entrance being flanked by low, square-topped towers. As a center for tourists, Inverness is increasingly popular and motor cars are very common. The roads of the surrounding country are generally excellent, and a trip of two hundred miles will take one to John O'Groats, the extreme northern point of Scotland. The country around has many spots of interest. Cawdor Castle, where tradition says Macbeth murdered Duncan, is on the Nairn road, and on the way to this one may also visit Culloden Moor, a grim, shelterless waste, where the adherents of Prince Charlie were defeated April 16th, 1746. This was the last battle fought on British soil, and the site is marked by a rude round tower built from stones gathered from the battlefield.

From Inverness an unsurpassed highway leads to Aberdeen, a distance of a little over one hundred miles. It passes through a beautiful country, the northeastern Scottish Lowlands, which looked as prosperous and productive as any section we saw. The smaller towns appeared much better than the average we had so far seen in Scotland; Nairn, Huntly, Forres, Keith and Elgin more resembling the better English towns of similar size than Scotch towns which we had previously passed through. At Elgin are the ruins of its once splendid cathedral, which in its best days easily ranked as the largest and most imposing church in Scotland. Time has dealt hardly with it, and the shattered fragments which remain are only enough to confirm the story of its magnificence. Fire, and vandals who tore the lead from the roof for loot having done their worst, the cathedral served the unsentimental Scots of the vicinity as a stone-quarry until recent years, but it is now owned by the crown and every precaution taken to arrest further decay.

The skies were lowering when we left Inverness and the latter half of the journey was made in the hardest rainstorm we encountered on our tour. We could not see ten yards ahead of us and the water poured down the hills in torrents, yet our car ran smoothly on, the fine macadam road being little affected by the deluge. The heavy rain ceased by the time we reached Inverurie, a gray, bleak-looking little town, closely following a winding street, but the view from the high bridge which we crossed just on leaving the place made full amends for the general ugliness of the village.

It would be hard to find anywhere a more beautiful city than Aberdeen, with her clean, massively built structures of native gray granite, thickly sprinkled with mica facets that make it fairly glitter in the sunlight. Everything seems to have been planned by the architect to produce the most pleasing effect, and careful note must have been taken of surroundings and location in fitting many of the public buildings into their niches. We saw few more imposing structures in Britain than the new postoffice at Aberdeen, and it was typical of the solidity and architectural magnificence of the Queen City of the North. But Aberdeen will be on the route of any tourist who goes to Northern Scotland, so I will not write of it here. It is a great motoring center, with finely built and well equipped garages.

TOWERS OF ELGIN CATHEDRAL, NORTH SCOTLAND.

DUNNOTTAR CASTLE, STONEHAVEN, NEAR ABERDEEN.

As originally planned we were to go southward from Aberdeen by the way of Braemar and Balmoral in the very heart of the Highland country—the route usually followed by British motorists. It passes through wild scenery, but the country has few historic attractions. The Motor Union representative had remarked that we should probably want to spend several days at Braemar, famous for its scenic surroundings—the wild and picturesque dales, lakes and hills near at hand; but to Americans, from the country of the Yellowstone and Yosemite, the scenery of Scotland can be only an incident in a tour. From this consideration, we preferred to take the coast road southward, which, though it passes through a comparatively tame-looking country, is thickly strewn with places replete with stirring and romantic incidents of Scottish history. Nor had we any cause to regret our choice.

Fifteen miles south of Aberdeen we came in sight of Dunnottar Castle, lying about two miles from the highway. We left the car by the roadside and followed the footpath through the fields. The ruin stands on a high, precipitous headland projecting far out into the ocean and cut off from the land side by a deep, irregular ravine, and the descent and ascent of the almost perpendicular sides was anything but an easy task. A single winding footpath leads to the grim old gateway, and we rang the bell many times before the custodian admitted us. Inside the gate the steep ascent continues through a rude, tunnellike passageway, its sides for a distance of one hundred feet or more pierced with many an embrasure for archers or musketeers. Emerging from this we came into the castle court, the center of the small plateau on the summit of the rock. Around us rose the broken, straggling walls, bare and bleak, without a shred of ivy or wall-flower to hide their grim nakedness. The place was typical of a rude, semi-barbarous age, an age of rapine, murder and ferocious cruelty, and its story is as terrific as one would anticipate from its forbidding aspect. Here it was the wont of robber barons to retire with their prisoners and loot; and later, on account of the inaccessibility, state and political prisoners were confined here from time to time. In the frightful "Whig's Vault," a semi-subterranean dungeon, one hundred and sixty covenanters—men and women—were for several months confined by orders of the infamous Claverhouse. A single tiny window looking out on the desolate ocean furnished the sole light and air for the great cavern, and the story of the suffering of the captives is too dreadful to tell here. The vault was ankle deep in mire and so crowded were the prisoners that no one could sit without leaning upon another. In desperation and at great risk, a few attempted to escape from the window, whence they clambered down the precipitous rock; but most of them were re-taken, and after frightful tortures were thrown into a second dungeon underneath the first, where light and air were almost wholly excluded. Such was Scotland in the reign of Charles Stuart II, and such a story seemed in keeping with the vast, dismal old fortress.

But Dunnottar, secluded and lonely as it was, did not escape the far-reaching arm of the Lord Protector, and in 1562 his cannon, planted on the height opposite the headland, soon brought the garrison to terms. It was known that the Scottish regalia—the crown believed to be the identical one worn by Bruce at his coronation, the jewelled scepter and the sword of state presented to James IV by the pope—had been taken for safety to Dunnottar, held in repute as the most impregnable stronghold in the North. The English maintained a close blockade by sea and land and were in strong hopes of securing the coveted relics. The story is that Mrs. Granger, the wife of a minister of a nearby village, who had been allowed by the English to visit the castle, on her departure carried the relics with her, concealed about her clothing. She passed through the English lines without interference, and the precious articles were safely disposed of by her husband, who buried them under the flagstones in his church at Kinneff, where they remained until the restoration of 1660. The English were intensely disappointed at the loss. The minister and his wife did not escape suspicion and were even subjected to torture, but they bravely refused to give information as to the whereabouts of the regalia.

We wandered about, following our rheumatic old guide, who pointed out the different apartments to us and, in Scotch so broad that we had to follow him very closely, told us the story of the fortress. From the windows everywhere was the placid, shimmering summer sea, its surface broken into silvery ripples by the fresh morning wind, but it was left to the imagination to conceive the awful desolation of Dunnottar Castle on a gray and stormy day. The old man conducted us to the keep, and I looked over a year's record in the visitors' book without finding a single American registered, and was more than ever impressed as to the manner in which the motor car will often bring the tourist from the States into a comparatively undiscovered country. The high tower of the keep, several hundred feet above the sea, afforded scope for a most magnificent outlook. One could get a full sweep of the bleak and sterile country through which we had passed, lying between Aberdeen and Stonehaven, and which Scott celebrated as the Muir of Drumthwacket. It was with a feeling of relief that we passed out of the forbidding portals into the fresh air of the pleasant July day, leaving the old custodian richer by a few shillings, to wonder that the "American Invasion" had reached this secluded old fortress on the wild headland washed by the German Ocean.

From Stonehaven we passed without special incident to Montrose, following an excellent but rather uninteresting road, though an occasional fishing-village and frequent view of the ocean broke the monotony of the flying miles. Montrose is an ancient town delightfully situated between the ocean and a great basin connected with the sea by a broad strait, over which a suspension bridge five hundred feet long carried us southward. I recall that it was at Montrose where an obliging garage man loaned me an "accumulator"—my batteries had been giving trouble—scouting the idea of a deposit, and I gave him no more than my agreement to return his property when I reached Edinburgh.

At Arbroath are the ruins of the most extensive of the Scotch abbeys, scanty indeed, but still enough to show its state and importance in the "days of faith." Here once reigned the good abbott celebrated by Southey in his ballad of Ralph the Rover, familiar to every schoolboy. Ten miles off the coast is the reef where

> "The abbott of Aberbrothok
> Had placed a bell on the Inchcape rock.
> Like a buoy in the storm it floated and swung,
> And over the waves its warning rung."

And where the pirate, out of pure malice, "To vex the abbott of Aberbrothok," cut the bell from its buoy only to be lost himself on the reef a year later. The abbey was founded by William the Lion in 1178, but war, fire and fanaticism have left it sadly fragmentary. Now it is the charge of the town, but the elements continue to war upon it and the brittle red sandstone of which it is built shows deeply the wear of the sea wind.

Dundee, no longer the "Bonnie Dundee" of the old ballad, is a great straggling manufacturing city, whose ancient landmarks have been almost swept away. Its churches are modern, its one remaining gateway of doubtful antiquity, and there is little in the city itself to detain the tourist. If its points of interest are too few to warrant a stay, its hotels—should the one given in the guide-book and also locally reputed to be the best, really merit this distinction—will hardly prove an attraction. It is a large, six-story building, fairly good-looking from the outside, but inside dirty and dilapidated, with ill-furnished and uncomfortable rooms. When we inquired of the manageress as to what might be of especial interest in Dundee, she considered awhile and finally suggested—the cemetery. From our hotel window we had a fine view of the broad estuary of the Tay with its great bridge, said to be the longest in the world. It recalled the previous Tay bridge, which fell in a storm in 1879, carrying down a train, from which not a single one of the seventy or more passengers escaped. Around Dundee is crowded much of historic Scotland, and many excursions worth the while may be made from the city by those whose time permits.

From Dundee an excellent road leads to Stirling by the way of Perth. There is no more beautiful section in Scotland than this, though its beauty is not the rugged scenery of the Highlands. Low hills, rising above the wooded valleys, with clear streams winding through them; unusually prosperous-looking farm-houses; and frequent historic ruins and places—all combine to make the forty or fifty miles a delightful drive. We did not pause at Perth, a city with a long line of traditions, nor at Dunblane, with its severely plain cathedral founded in 1100 but recently restored.

Stirling, the ancient capital, with its famous castle, its memories of early kings, of Wallace, Bruce and of Mary Stuart, and with its wonderfully beautiful and historic surroundings, is perhaps the most interesting town of Scotland. No one who pretends to see Scotland will miss it, and no motor tour worthy of the name could be planned that would not lead through the quaint old streets. From afar one catches a glimpse of the castle, perched, like that of Edinburgh, on a mighty rock, rising almost sheer from a delightfully diversified plain. It is a many-towered structure, piercing the blue sky and surrounded by an air of sullen inaccessibility, while the red-cross flag flying above it proclaims it a station of the king's army. It is not by any means the castle of the days of Bruce and Wallace, having been rebuilt and adapted to the purpose of military barracks. True, many of the ancient portions remain, but the long, laborious climb to the summit of the rock and the battlements of the castle will, if the day be fine, be better repaid by the magnificent prospect than by anything else. If the barrack castle is a little disappointing, the wide sweep of country fading away into the blue mountains on the west—-Ben

Venue, Ben Ledi and Ben Lomond of "The Lady of the Lake"—eastward the rich lowlands, running for miles and miles down the fertile valley of the Forth, dotted with many towns and villages; the wooded hills to the north with the massive tower of the Wallace monument and the dim outlines of the ruins of Cambuskenneth Abbey; or, near at hand, the old town under your very eye and the historic field of Bannockburn just adjoining, will make ample amends. The story of "The Lady of the Lake" pictures Stirling in its palmiest days, and no one who visits the castle will forget the brilliant closing scene of the poem. Here too,

> "The rose of Stuart's line
> Has left the fragrance of her name,"

for Mary was hurried for safety to the castle a few days after her birth at Linlithgow Palace, and as a mere baby was crowned Queen of Scotland in the chapel. The parish church was also the scene of many coronations, and in the case of James VI, later James I of England, John Knox preached the sermon.

One cannot go far in Scotland without crossing the path of Prince Charlie or standing in the shadow of some ancient building associated with the melancholy memory of Queen Mary, and, despite the unquestioned loyalty of the Scottish people to the present government, there seems to linger everywhere a spirit of regret over the failure of the chevalier to regain the throne of his fathers. Perhaps it is scarcely expressed—only some word dropped in casual conversation, some flash of pride as you are pointed to the spots where Prince Charlie's triumphs were won, or some thinly veiled sentiment in local guide-books will make it clear to you that Scotland still cherishes the memory of the prince for whom her fathers suffered so much. Passing Falkirk, now a large manufacturing town, dingy with the smoke from its great furnaces, we were reminded that near here in 1746 the prince gained one of his most decisive victories, the precursor of the capture of Edinburgh by his army. A few miles farther on is Linlithgow with its famous palace, the birthplace of the Queen of Scots. This more accords with our idea of a royal residence than the fortified castles, for it evidently was never intended as a defensive fortress. It stands on the margin of a lovely lake, and considering its delightful situation and its comparative comfort, it is not strange that it was a favorite residence of the Scottish kings. It owes its dismantled condition to the wanton spite of the English dragoons, who, when they retreated from Linlithgow in face of the Highland army in 1746, left the palace in flames.

From Linlithgow the broad highway led us directly into Edinburgh by the way of Princess Street.

XI

FROM EDINBURGH TO YORKSHIRE

Two men above all others and everything else are responsible for the romantic fame which the bleak and largely barren Land of Scots enjoys the English-speaking world over. If Robert Burns and Walter Scott had never told the tales and sung the songs of their native land, no endless streams of pilgrims would pour to its shrines and its history and traditions would be vastly second in interest to those of England and Wales. But the Wizard of the North touched Scotia's rough hills with the rosy hues of his romance. He threw the glamour of his story around its crumbling ruins. Through the magic of his facile pen, its petty chiefs and marauding nobles assumed heroic mould and its kings and queens—rulers over a mere handful of turbulent people—were awakened into a majestic reality. Who would care aught for Prince Charlie or his horde of beggarly Highlanders were it not for the song of Burns and the story of Scott? Nor would the melancholy fate of Queen Mary have been brought so vividly before the world—but wherefore multiply instances to illustrate an admitted fact?

In Edinburgh we were near the center from which Scott's vast influences radiated. The traditions of Burns overshadowed Southwestern Scotland and the memories of Scott seem to be indentified with the cities, the villages, the solitary ruins, the hills and vales of the eastern coast. We note as we pass along Princess Street, one of the finest thoroughfares in Britain, the magnificent monument to the great author—the most majestic tribute ever erected to a literary man—a graceful Gothic spire, towering two hundred feet into the sky. The city is full of his memories. Here are many of the places he celebrated in his stories, his haunts for years, and the house where he retired after financial disaster to face a self-chosen battle with a gigantic debt which he might easily have evaded by a mere figment of the law.

However, one can hardly afford to take from a motor tour the time which should rightly be given to Edinburgh, for the many attractions of the Athens of the North might well occupy a solid week. Fortunately, a previous visit by rail two years before had solved the problem for us and we were fairly familiar with the more salient features of the city. There is one side-trip that no one should miss, and though we had once journeyed by railway train to Melrose Abbey and Abbottsford House, we could not forego a second visit to these famous shrines and to Dryburgh Abbey, which we had missed before. Thus again we had the opportunity of contrasting the motor car and the railway train. I remembered distinctly our former trip to Melrose by rail. It was on a Saturday afternoon holiday when crowds of trippers were leaving the city, packed in the uncomfortable compartments like sardines in a box—not one in a dozen having a chance to sit. We were driven from Melrose to Abbottsford House at a snail's pace, consuming so much time that a trip to Dryburgh Abbey was out of the question, though we had left Edinburgh about noon. By motor, we were out of the city about three o'clock, and though we covered more than eighty miles, we were back before lamp-lighting time. The road to Dryburgh Abbey runs nearly due south from Edinburgh, and the country through which we passed was hardly so prosperous looking as the northeastern section of Scotland—much of it rather rough-looking country, adapted only for sheep-grazing and appearing as if it might be reclaimed moorland.

The tomb of Walter Scott is in Dryburgh Abbey, and with the possible exception of Melrose it probably has more visitors than any other point in Scotland outside of Edinburgh. The tourist season had hardly begun, yet the caretaker told us that more than seventy people had been there during the day and most of them were Americans. The abbey lies on the margin of the River Tweed, the silver stream so beloved of Scott, and though sadly fragmentary, is most religiously cared for and the decay of time and weather held in check by constant repairs and restoration. The many thousands of admission fees every year no doubt form a fund which will keep this good work going indefinitely. The weather-beaten walls and arches were overgrown with masses of ivy and the thick, green grass of the newly mown lawn spread beneath like a velvet carpet. We had reached the ruin so late that it was quite deserted, and we felt the spirit of the place all the more as we wandered about in the evening silence. Scott's tomb, that of his wife and their eldest son are in one of the chapels whose vaulted roof still remains in position. Tall iron gates between the arches enclose the graves, which are marked with massive sarcophagi of Scotch granite. Dryburgh Abbey was at one time the property of the Scott family, which accounts for its use as their burial-ground. It has passed into other hands, but interments are still made on rare occasions. The spot was one which always interested and delighted Scott and it was his expressed wish that he be buried there.

We had been warned that the byways leading to the abbey from the north of the Tweed were not very practicable for motors and we therefore approached it from the other side. This made it necessary to cross the river on a flimsy suspension bridge for foot-passengers only, and a notice at each end peremptorily forbade that more than half a dozen people pass over the bridge at one time. After crossing the river it was a walk of more than a mile to the abbey, and as we were tempted to linger rather long it was well after six o'clock when we re-crossed the river and resumed our journey. Melrose is twelve miles farther on and the road crosses a series of rather sharp hills. We paused for a second glimpse of Melrose Abbey, which has frequently been styled the most perfect and beautiful ecclesiastical ruin in Britain. We were of the opinion, however, that we had seen at least three or four others more extensive and of greater architectural merit. Undoubtedly the high praise given Melrose is due to the fame which it acquired from the poems and stories of Scott. The thousands of pilgrims who come every year are attracted by this alone, since the abbey had no extraordinary history and no tomb of king or hero is to be found in its precincts. Were it not for the weird interest which the

"Lay of the Last Minstrel" has thrown around Melrose, its fame would probably be no greater than that of the abbeys of Jedburgh and Kelso in the same neighborhood. Abbottsford House is only three miles from Melrose, but it is closed to visitors after five o'clock and we missed a second visit, which we should have liked very much. Upon such things the motorist must fully inform himself or he is liable to many disappointments by reaching his objective point at the wrong time.

We returned to Edinburgh by the way of Galashiels, a manufacturing town of considerable size that lay in a deep valley far below the road which we were following along the edges of the wooded hills. This road abounded in dangerous turns and caution was necessary when rounding sharp curves that, in places, almost described a circle. We had a clear right-of-way, however, and reached Edinburgh before nine o'clock. A delightful feature of summer touring in Britain is the long evening, which is often the pleasantest time for traveling. The highways are usually quite deserted and the mellow effect of the sunsets and the long twilights often lend an additional charm to the landscapes. In the months of July and August in Scotland daylight does not begin to fade away until from nine to ten, and in northern sections the dawn begins as early as two or three o'clock. During our entire tour we found it necessary to light our lamps only two or three times, although we were often on the road after nine o'clock. Though Edinburgh has unusually broad and well paved streets, it is a trying place for a motorist. The people make little effort to keep to the sidewalk, but let the fellow who is driving the car do the looking out for them. In no city through which we passed did I find greater care necessary. Despite all this, accidents are rare, owing to the fact that drivers of motor cars in Great Britain have had the lesson of carefulness impressed upon them by strict and prompt enforcement of police regulations.

We left Edinburgh the next forenoon with a view of making Berwick-on-Tweed our stopping place for the evening—not a long distance in miles but a considerable one measured in spots of historical importance. The road much of the way skirts the ocean and is a magnificent highway leading through a number of quaint towns famous in Scotch song and story. Numerous battlefields are scattered along the way, but we found it difficult to locate a battlefield when we passed it, and generally quit trying. In fact, in the days of border warfare the whole south of Scotland was the scene of almost continuous strife, and battles of greater or less importance were fought everywhere with the English in the centuries of fierce hatred which existed between the two nations. The Scots held their own wonderfully well, considering their greatly inferior numbers and the general poverty of their country. The union, after all, was brought about not by conquest but by a Scotch king going to London to assume the crown of the two kingdoms. The famous old town of Berwick-on-Tweed bore the brunt of the incursions from both sides on the eastern coast, as did Carlisle on the west. The town of Dunbar, situated on the coast about midway between Edinburgh and Berwick, was of great importance in border history. It had an extensive and strongly fortified castle, situated on the margin of a cliff overhanging the ocean, and which was for a time the residence of Queen Mary after her marriage with Darnley. Nothing now remains of this great structure save a few crumbling walls of red sandstone, which are carefully propped up and kept in the best possible repair by the citizens, who have at last come to realize the cash value of such a ruin. If such a realization had only come a hundred years ago, a great service would have been done the historian and the antiquarian. But this is no less true of a thousand other towns than of Dunbar. No quainter edifice did we see in all Britain than Dunbar's Fifteenth Century town hall. It seemed more characteristic of an old German town than of Scotland. This odd old building is still the seat of the city government.

TOWN HOUSE, DUNBAR, SCOTLAND.

Our route from Dunbar ran for a long way between the hills of Lammermoor and the ocean and abounded in delightful and striking scenery. We were forcibly reminded of Scott's mournful story, "The Bride of Lammermoor," as we passed among the familiar scenes mentioned in the book, and it was the influence of this romantic tale that led us from the main road into narrow byways and sleepy little coast towns innocent of modern progress and undisturbed by the rattle of railways trains. No great distance from Berwick and directly on the ocean stands Fast Castle, said to be the prototype of the Wolf's Crag of "Lammermoor." This wild story had always interested me in my boyhood days and for years I had dreamed of the possibility of some time seeing the supposed retreat of the melancholy Master of Ravenswood. We had great difficulty in locating the castle, none of the people seeming to know anything about it, and we wandered many miles among the hills through narrow, unmarked byways, with little idea of where we were really going. At last, after dint of inquiry, we came upon a group of houses which we were informed were the headquarters of a large farm of about two thousand acres, and practically all the people who worked on the farm lived, with their families, in these houses. The superintendent knew of Fast Castle, which he said was in a lonely and inaccessible spot, situated on a high, broken headland overlooking the ocean. It was two or three miles distant and the road would hardly admit of taking the car any farther. He did not think the ruin was worth going to see, anyhow; it had been cared for by no one and within his memory the walls had fallen in and crumbled away. Either his remarks or the few miles walk discouraged me, and after having traveled fully thirty miles to find this castle, I turned about and went on without going to the place at all, and of course I now regret it as much as anything I failed to do on our whole tour. I shall have to go to Fast Castle yet—by motor car.

After regaining the main road, it was only a short run along the edge of the ocean to Berwick-on-Tweed, which we reached early in the evening. I recall no more delightful day during our tour. It had been fresh and cool, and the sky was perfectly clear. For a great part of the way the road had passed within view of the ocean, whose deep unruffled blue, entirely unobscured by the mists which so often hang over the northern seas, stretched away until it was lost in the pale, sapphire hues of the skies. The country itself was fresh and bright after abundant rains, and as haymaking was in progress in many places along the road, the air was laden with the scent of the newly mown grasses. Altogether, it was a day long to be remembered.

Berwick-on-Tweed lies partly in England and partly in Scotland, the river which runs through it forming the boundary line. An odd bridge built by James I connects the two parts of the town, the highest point of its archway being nearest the Scottish shore and giving the effect of "having its middle at one end," as some Scotch wit has expressed it. The town was once strongly fortified, especially on the Scottish side, and a castle was built on a hill commanding the place. Traces of the wall surrounding the older part of the city still remain; it is easy to follow it throughout its entire course. When the long years of border warfare ended, a century and a half ago, the town inside of the wall must have appeared much the same as it does today. It is a town of crooked streets and quaint buildings, set down without the slightest reference to the points of the compass. The site of the castle is occupied by the railway station, though a few crumbling walls of the former structure still remain. The station itself is now called The Castle and reproduces on a smaller scale some of the architectural features of the ancient fortress.

We started southward from Berwick the following morning over the fine road leading through Northumberland. About ten miles off this road, and reached by narrow byways, is the pleasant little seacoast village of Bamborough, and the fame of its castle tempted us to visit it. I had often wondered why some of the old-time castles were not restored to their pristine magnificence—what we should have if Kenilworth or Raglan were re-built and to their ancient glory there were added all the modern conveniences for comfort. I found in Bamborough Castle a case exactly to the point. Lord Armstrong, the millionaire shipbuilder, had purchased this castle—almost a complete ruin—and when he began restoration only the Norman tower of the keep was intact; and besides this there was little except the foundation walls. Lord Armstrong entirely rebuilt the castle, following the original plan and designs, and the result is one of the most striking and pleasing of the palatial residences in England. The situation, on a high headland extending into the ocean, commands a view in every direction and completely dominates the sleepy little village lying just beneath. The castle is of great antiquity, the records showing that a fortress had been built on this side in the Fifth Century by Ida, King of Northumberland, though the present building largely reproduces the features of the one founded in the time of the Conqueror.

Lord Armstrong died the year before the work on the castle was completed and it passed into the hands of his nephew. It is open to visitors only one day in the week, and it happened, as usual, that we had arrived on the wrong day. Fortunately, the family were absent, and our plea that we were Americans who had come a long distance to see the place was quite as effective here as in other cases. The housekeeper showed us the palace in detail that we could hardly have hoped for under other circumstances. The interior is fitted in the richest and most magnificent style, and I have never seen the natural beauties of woodwork brought out with better effect. How closely the old-time construction was followed in the restoration is shown by the fact that the great open roof of the banqueting hall is put together with wooden pins, no nail having been used. The castle has every modern convenience, even hot-water heating—a rare thing in England—being installed. When we saw what an excellent result had been attained in the restoration, we could not but wonder that such a thing has not oftener been done. In the village churchyard is the massive gray granite monument erected to the memory of Grace Darling, who lived and died in Bamborough, and a brass tablet in the ancient church is inscribed with the record of her heroism. The lighthouse which was kept by her father is just off Bamborough Head, and it was from this, in the face of a raging storm, that she launched her frail boat and saved several people from a foundering ship. Only four years later she succumbed to consumption, but her unparalleled bravery has made the name of this young girl a household word wherever the English language is spoken.

BAMBOROUGH CASTLE, NORTHUMBERLAND.

On leaving Bamborough we came as nearly getting lost in the narrow, winding byways as at any time during our tour. A bridge under repair on the direct route to the main road compelled us to resort to byways which were unmarked by signboards and in as ill condition as many American roads. Nor could the people of whom we inquired give us intelligent direction. We finally reached the road again after a loss of an hour or more.

A short time afterwards we came to Alnwick, whose castle is one of the most extensive and complete specimens of mediaeval architecture in England. In the last century it has been largely restored, following out the original design of the exterior, at least, and is now the residence of the Duke of Northumberland. Usually it is open to visitors, but in the confusion that followed the visit of the king the day before, the castle and its great park had been closed until the next week. We had seen the interior of so many similar places that this was not so much of a disappointment, especially as we had a splendid view of the old fortress from the outside and also from the courtyard. On the battlements of this castle are numerous stone figures of men in the act of hurling down missiles on the heads of foes who might besiege it. This was quite common in early days and feudal barons perhaps thought to make up for their shortage of real men by placing these effigies on the walls of their fortresses, but Alnwick is the only castle on which the figures still remain. The town itself was still in holiday attire in honor of its royal guest of the preceding day. The buildings were covered with the national colors and many decorations and illuminations had been planned to celebrate the occasion. Alnwick is one of the most typical of the English feudal towns. It is owned largely by the Duke of Northumberland, who appears to be popular with his tenantry, the latter having erected, in honor of their noble landlord, a lofty column surmounted by the figure of a lion. Every view from the distance for miles around is dominated by the battlemented and many-towered walls of the castle, which surmounts a hill overlooking the town. The story of Alnwick and its castle would be long to tell, for they bore the brunt of many Scotch incursions and suffered much at the hands of the fierce marauders from the north.

Our afternoon's run led us from Alnwick to Durham, passing through Newcastle-on-Tyne. Newcastle is a large commercial city, famous for its mining and shipbuilding industries, and has but little to engage the attention of the tourist. Our pause was a short one, and we reached Durham in good time after a run of over one hundred miles, broken by several lengthy stops on the way.

The main street of Durham in many places is barely wide enough for two vehicles to pass. It winds and twists through the town in such a way that one seems to be almost moving in a circle at times and constant inquiry is necessary to keep from being lost on the main street of a city of fifteen or twenty thousand. The town is almost as much of a jumble as if its red, tile-roof buildings had been promiscuously thrown to their places from Cathedral Hill. Durham is strictly an ecclesiastical center. There is little except the cathedral, which, in addition to being one of the most imposing, occupies perhaps the finest site of any of the great English churches. Together with Durham Castle, it monopolizes the summit of a hill which at its base is three-quarters surrounded by the river. The greater part of the cathedral dates back seven or eight hundred years, but additions have been made from time to time so that nearly all styles of architecture are represented. Tradition has it that it was founded by St. Cuthbert, whose chief characteristic is declared to have been his antipathy toward women of all degrees. A curious relic of this peculiarity of the saint remains in a granite cross set in the center of the floor of the nave, beyond which, in the earlier days, no woman was ever allowed to pass. The interior of the church is mainly in the massive and imposing Norman style. The carved stone screen is one of the most elaborate and perfect in Britain, and dates back from the Thirteenth Century. The verger told us of the extreme care which must be taken to preserve this relic. He said that the stone of the screen is rather soft and brittle, and that in cleaning it was never touched, the dust being blown away with bellows. Durham, in common with most of the cathedrals, suffered severely at the hands of the Parliamentarians under Cromwell. It was used as a prison for a part of the Scotch army captured at the battle of Dunbar, and as these Presbyterians had almost as much contempt for images as the Cromwellians themselves, many of the beautiful monuments in the cathedral were broken up. Durham, like Canterbury, is a town that is much favored by the artists, and deservedly so. The old buildings lining the winding river and canal form in many places delightful vistas in soft colors almost as picturesque as bits of Venice itself. The hotels, however, are far from first-class, and one would probably be more comfortable at Newcastle. Speaking of hotels, we did not at any time engage accommodations in advance, and Durham was the only town where we found the principal hotel with all rooms taken. With the rapid increase of motoring, however, it will probably become necessary to telegraph for accommodations at the best hotels. And telegraphing is an exceedingly easy thing in England. A message can be sent from any postoffice at a cost of sixpence for the first ten words.

XII

IN OLD YORKSHIRE

York is by far the largest of the English shires, a widely diversified country, ranging from fertile farm land to broken hills and waste moorland, while its river valleys and considerable coast line present greatly varied but always picturesque scenery. The poet describes the charms of Yorkshire as yielding

> "Variety without end, sweet interchange
> Of hill and valley, river, wood and plain."

Nor did we find this description at all inapt as we drove over its excellent roads during the fine July weather. But the Yorkshire country is doubly interesting, for if the landscape is of surpassing beauty, the cities, the villages, the castles and abbeys, and the fields where some of the fiercest battles in Britain have been fought, have intertwined their associations with every hill and valley. Not only the size of the shire, but its position—midway between London and the Scottish border, and extending almost from coast to coast—made it a bulwark, as it were, against the incursions of the Scots and their numerous sympathizers in the extreme north of England. No part of England is more thickly strewn with attractions for the American tourist and in no other section do conditions for motor travel average better.

From London to York, the capital city of the shire, runs the Great North Road, undoubtedly the finest highway in all Britain. It is laid out on a liberal scale, magnificently surfaced and bordered much of the way by wide and beautifully kept lawns and at times skirted with majestic trees. We saw a facsimile of a broadside poster issued about a century ago announcing that the new lightning coach service installed on this road between London and York would carry passengers the distance of one hundred and eighty-eight miles in the astonishingly short space of four days. This coach, of course, traveled by relays, and at what was then considered breakneck speed. Over this same highway it would now be an easy feat for a powerful car to cover the distance in three or four hours. The great North Road was originally constructed by the Romans to maintain the quickest possible communication between London and Eboracum, as York was styled during the Roman occupation.

The limitation of our time had become such that we could but feel that our tour through Yorkshire must be of the most superficial kind. Not less than two weeks of motoring might well be spent in the county and every day be full of genuine enjoyment. The main roads are among the best in England and afford access to most of the important points. We learned, however, that there is much of interest to be reached only from byways, but that these may lead over steep and even dangerous hills and are often in not much better condition than our American roads.

We left Durham about noon, following a rather indirect route to Darlington; from thence, through hawthorne-bordered byways, we came to Richmond, one of the quaintest and most representative of the old Yorkshire towns. We happened here on market day and the town was crowded with farmers from the surrounding country. Here we saw many types of the Yorkshire man, famed for his shrewdness and fondness for what we would call "dickering." Much of the buying and selling in English towns is done on market day; live stock, produce, farm implements, and almost every kind of merchandise are sold at auction in the public market place. If a farmer wants to dispose of a horse or to buy a mowing machine, he avails himself of this auction and the services of a professional auctioneer. Such an individual was busily plying his vocation in front of the King's Head Hotel, and the roars of laughter from the farmers which greeted his sallies as he cried his wares certainly seemed to indicate that the charge that Englishmen can not appreciate humor—at least of a certain kind—is a base slander. As Richmond is the center of one of the best farming districts in Yorkshire, its market day was no doubt a typical one.

Richmond Castle at one time was one of the most formidable and strongly situated of the northern fortresses. It stands on an almost perpendicular rock, rising one hundred feet above the River Swale, but with the exception of the Norman keep the ruins are scanty indeed. There is enough of the enclosing walls to give some idea of the extent of the original castle, which covered five acres, its magnificent position commanding the whole of the surrounding country. The keep is now used as a military storehouse. The soldier-guard in charge was very courteous and relieved us the necessity of securing a pass from the commandant, as was required by a notice at the castle entrance. He conducted us to the top of the great tower, from which we were favored with one of the finest views in Central England and one that is almost unobstructed in every direction. Unfortunately, a blue mist obscured much of the landscape, but the guard told us that on clear days York Minster, more than forty miles away, could be easily seen. Near at hand, nestling in the valley of the Swale, are the ivy-covered ruins of Easby Abbey; while still nearer, on the hillside, the great tower of Grey Friars Church is all that remains of another once extensive monastery. In no way can one get a more adequate idea of the park-like beauty of the English landscape than to view it from such point of vantage as the keep of Richmond Castle. Richmond Church is an imposing structure standing near the castle and has recently been restored as nearly as possible to its ancient state. An odd feature of the church is the little shop built in the base of the tower, where a tobacconist now plies his trade.

From the castle tower, looking down the luxuriant valley, we noticed at no great distance, half hidden by the trees, the outlines of a ruined church—the Easby Abbey which I have just mentioned as one of the numerous Yorkshire ruins. It is but a few furlongs off the road by which we left Richmond and the byway we entered dropped down a sharp hill to the pleasant spot on the riverside, where the abbey stands. The location is a rather secluded one and the painstaking care noticeable about so many ruins is lacking. It is surrounded by trees, and a large elm growing in the very midst of the walls and arches flung a network of sun and shade over the crumbling stones. The murmur of the nearby Swale and the notes of the English thrushes filled the air with soft melody. Amid such surroundings, we hardly heard the old custodian as he pointed out the different apartments and told us the story of the palmy days of the abbey and of its final doom at the relentless hands of Henry VIII. Near by is a tiny church, which no doubt had served the people of the neighborhood as a place of worship since the abbey fell into ruin.

The day, which had so far been fine, soon began to turn cold—one of those sudden and disagreeable changes that come in England and Scotland in the very midst of summertime, an experience that happens so often that one can not wonder at Byron's complaint of the English winter, "closing in July to re-commence in August." At no time in the summer were we able to dispense for any length of time with heavy wraps and robes while on the road. From Richmond we hastened away over a fine and nearly straight road to Ripon, whose chief attraction is its cathedral. Speaking of cathedrals again, I might remark that our tour took us to every one of these, with one exception—in England and Scotland, about thirty in all—and the exception, Beverly Minster, is but newly created and relatively of lesser importance.

Ripon is one of the smaller cathedrals and of less importance in historical associations. It occupies a magnificent site, crowning a hill rising in the very center of the town, and from a distance gives the impression of being larger than it really is. It presents a somewhat unfinished aspect with its three low, square-topped towers, once surmounted by great wooden spires, which became unsafe and were taken down, never to be replaced. These must have added wonderfully to the dignity and proper proportion of the church.

Just outside Ripon lies Fountains Abbey, undoubtedly the most striking and best preserved ecclesiastical ruin in England. It is on the estate of the Marquis of Ripon, adjoining the town, and this nobleman takes great pride in the preservation of the abbey. The great park, which also surrounds his residence, is thrown open every day and one has full liberty to go about it at pleasure. It is a popular resort, and on the day of our visit the number of people passing through the gate exceeded five hundred. The gatekeeper assured us that a thousand visitors on a single day was not an uncommon occurrence. The abbey stands in a wooded valley on the margin of a charming little river, and underneath and around the ruin is a lawn whose green loveliness is such as can be found in England alone. There is no room in this record for the description of such a well known place or for its story. The one feature which impressed us most, and which is one of the finest specimens of Norman architecture in England, is the great cellarium, where the monks stored their wine in the good old days. The vaulted roof of this vast apartment, several hundred feet in length, is in perfect condition and shows how substantially the structure must have been built Fountains Abbey shared the fate of its contemporaries at the hand of Henry VIII, who drove the monks from its shelter, confiscating their property and revenues. It was growing late when we left Ripon for York, but the road was perfect and we had no trouble in covering the twenty miles or more in about an hour. We were soon made comfortable at the Station Hotel in

York, one of the oldest and most interesting of the larger cities.

The following day being Sunday, we availed ourselves of the opportunity of attending services at the Minster. The splendid music of the great organ was enough to atone for the long dreary chant of the litany, and the glory of the ancient windows, breaking the gloom of the church with a thousand shafts of softened light, was in itself an inspiration more than any sermon—at least to us, to whom these things had the charm of the unusual.

York Minster, with the exception of St. Paul's in London, is the largest cathedral in England and contests with Canterbury for first place in ecclesiastical importance. Its greatest glory is its windows, which are by far the finest of any in England. Many of them date back to the Thirteenth and Fourteenth Centuries, and when one contemplates their subdued beauty it is easy to understand why stained-glass making is now reckoned one of the lost arts. These windows escaped numerous vicissitudes which imperiled the cathedral, among them the disastrous fires which nearly destroyed it on two occasions within the last century. The most remarkable of them all is the "Five Sisters" at the end of the nave, a group of five slender, softly-toned windows of imposing height. The numerous monuments scattered throughout the church are of little interest to the American visitor. We were surprised at the small audiences which we found at the cathedrals where we attended services. A mere corner is large enough to care for the congregations, the vast body of the church being seldom used except on state occasions. Though York is a city of seventy-five thousand population, I think there were not more than four or five hundred people in attendance, though the day was exceptionally fine.

There are numerous places within easy reach of York which one should not miss. A sixty-mile trip during three or four hours of the afternoon gave us the opportunity of seeing two abbey ruins, Helmsley Castle and Laurence Sterne's cottage at Coxwold. Our route led over a series of steep hills almost due north to Helmsley, a town with unbroken traditions from the time of the Conqueror. Its ancient castle surrendered to Fairfax with the agreement that "it be absolutely demolished and that no garrison hereafter be kept by either party." So well was this provision carried out that only a ragged fragment remains of the once impregnable fortress, which has an added interest from its connection with Scott's story, "The Fortunes of Nigel"

Two miles from Helmsley is Rievaulx Abbey, situated in a deep, secluded valley, and the narrow byway leading to the ruin was so steep and rough that we left the car and walked down the hill. A small village nestles in the valley, a quiet, out-of-the-way little place whose thatched cottages were surrounded by a riot of old-fashioned flowers and their walls dashed with the rich color of the bloom-laden rose vines. Back of the village, in lonely grandeur, stands the abbey, still imposing despite decay and neglect. Just in front of it is the cottage of the old custodian, who seemed considerably troubled by our application to visit the ruins. He said that the place was not open on Sunday and gave us to understand that he had conscientious scruples against admitting anyone on that day. The hint of a fee overcame his scruples to such an extent that he intimated that the gates were not locked anyway and if we desired to go through them he did not know of anything that would prevent us. We wandered about in the shadows of the high but crumbling walls, whose extent gave a strong impression of the original glory of the place, and one may well believe the statement that, at the time of the Dissolution, Rievaulx was one of the largest as well as richest of the English abbeys. The old keeper was awaiting us at the gateway and his conscientious scruples were again awakened when we asked him for a few post-card pictures. He amiably intimated his own willingness to accommodate us, but said he was afraid that the "old woman" (his wife) wouldn't allow it, but he would find out. He returned after a short interview in the cottage and said that there were some pictures on a table in the front room and if we would go in and select what we wanted and leave the money for them it would be all right.

OLD COTTAGES AT COCKINGTON.

On our return from Helmsley, we noticed a byway leading across the moorland with a sign-board pointing the way "to Coxwold." We were reminded that in this out-of-the-way village Laurence Sterne, "the father of the English novel," had lived many years and that his cottage and church might still be seen. A narrow road led sharply from the beautiful Yorkshire farm lands, through which we had been traveling, its fields almost ready for the harvest, into a lonely moor almost as brown and bare as our own western sagebrush country. It was on this unfrequented road that we encountered the most dangerous hill we passed over during our trip, and the road descending it was a reminder of some of the worst in our native country. They called it "the bank," and the story of its terrors to motorists, told us by a Helmsley villager, was in no wise an exaggeration. It illustrates the risk often attending a digression into byroads not listed in the road-book, for England is a country of many hilly sections. I had read only a few days before of the wreck of a large car in Derbyshire where the driver lost control of his machine on a gradient of one in three. The car dashed over the embankment, demolishing many yards of stone wall and coming to rest in a valley hundreds of feet beneath. And this was only one of several similar cases. Fortunately, we had only the descent to make. The bank dropped off the edge of the moorland into a lovely and fertile valley, where, quite unexpectedly, we came upon Bylands Abbey, the rival of Rievaulx, but far more fallen into decay. It stood alone in the midst of the wide valley; no caretaker hindered our steps to its precincts and no effort had been made to prop its crumbling walls or to stay the green ruin creeping over it. The fragment of its great eastern window, still standing, was its most imposing feature and showed that it had been a church of no mean architectural pretension. The locality, it would seem, was well supplied with abbeys, for Rievaulx is less than ten miles away, but we learned that Bylands was founded by monks from the former brotherhood and also from Furness Abbey in Lancashire. In the good old days it seems to have been a common thing when the monks became dissatisfied with the establishment to which they were attached for the dissenters to start a rival abbey just over the way.

Coxwold is a sleepy village undisturbed by modern progress, its thatched cottages straggling up the crooked street that leads to the hilltop, crowned by the hoary church whose tall, massive octagonal tower dominates the surrounding country. It seems out of all proportion to the poverty-stricken, ragged-looking little village on the hillside, but this is not at all an uncommon impression one will have of the churches in small English towns. Across the road from the church is the old-time vicarage, reposing in the shade of towering elms, and we found no difficulty whatever in gaining admission to "Shandy Hall," as it is now called. We were shown the little room not more than nine feet square where Sterne, when vicar, wrote his greatest book, "Tristram Shandy." The kitchen is still in its original condition, with its rough-beamed ceiling and huge fireplace. Like most English cottages, the walls were covered with climbing roses and creepers and there was the usual flower-garden in the rear. The tenants were evidently used to visitors, and though they refused any gratuity, our attention was called to a box near the door which was labeled, "For the benefit of Wesleyan Missions."

Two or three miles through the byways after leaving Coxwold brought us into the main road leading into York. This seemed such an ideal place for a police trap that we traveled at a very moderate speed, meeting numerous motorists on the way. The day had been a magnificent one, enabling us to see the Yorkshire country at its best. It had been delightfully cool and clear, and lovelier views than we had seen from many of the upland roads would be hard to imagine. The fields of yellow grain, nearly ready for harvesting, richly contrasted with the prevailing bright green of the hills and valleys. Altogether, it was a day among a thousand, and in no possible way could one have enjoyed it so greatly as from the motor car, which dashed along, slowed up, or stopped altogether, as the varied scenery happened to especially please us.

York abounds in historic relics, odd corners and interesting places. The city was surrounded by a strong wall built originally by Edward I, and one may follow it throughout its entire course of more than two miles. It is not nearly so complete as the famous Chester wall, but it encloses a larger area. It shows to even a greater extent the careful work of the restorer, as do the numerous gate-towers, or "bars," which one meets in following the wall. The best exterior views of the minster may be had from vantage points on this wall, and a leisurely tour of its entire length is well worth while. The best preserved of the gate-towers is Micklegate Bar, from which, in the War of the Roses, the head of the Duke of York was exhibited to dismay his adherents. There were originally forty of these towers, of which several still exist. Aside from its world-famous minster, York teems with objects and places of curious and archaeological interest. There are many fine old churches and much mediaeval architecture. In a public park fragments still remain of St. Mary's Abbey, a once magnificent establishment, destroyed during the Parliamentary wars; but it must be said to the everlasting credit of the Parliamentarians that their commanders spared no effort to protect the minster, which accounts largely for its excellent preservation. The Commander-in-Chief, General Fairfax, was a native of Yorkshire and no doubt had a kindly feeling for the great cathedral, which led him to exert his influence against its spoliation. Such buildings can stand several fires without much damage, since there is little to burn except the roof, and the cathedrals suffered most severely at the hands of the various contending factions into which they fell during the civil wars.

The quaintest of old-time York streets is The Shambles, a narrow lane paved with cobblestones and only wide enough to permit the passing of one vehicle at a time. It is lined on either side with queer, half-timbered houses, and in one or two places these have sagged to such an extent that their tops are not more than two or three feet apart. In fact it is said that neighbors in two adjoining buildings may shake hands across the street. The Shambles no doubt took its name from the unattractive row of butcher shops which still occupy most of the small store-rooms on either side. Hardly less picturesque than The Shambles is the Petergate, and no more typical bits of old-time England may be found anywhere than these two ancient lanes. Glimpses of the cathedral towers through the rows of odd buildings is a favorite theme with the artists. Aside from its antiquity, its old-world streets and historic buildings are quite up to the best of the English cities. It is an important trading and manufacturing point, though the prophecy of the old saw,

> "Lincoln was, London is, York shall be.
> The greatest city of the three,"

seems hardly likely to be realized.

XIII

A ZIG-ZAG TRIP FROM YORK TO NORWICH

Late in the afternoon we left York over the Great North Road for Retford, from whence we expected to make the "Dukeries" circuit. The road runs through a beautiful section and passes many of the finest of the English country estates. It leads through Doncaster, noted for its magnificent church, and Bawtry, from whence came many of the Pilgrim Fathers who sailed in the Mayflower. This road is almost level throughout, and although it rained continuously, the run of fifty miles was made in record time—that is, as we reckoned record time.

At Retford we were comfortably housed at the White Hart Hotel, a well conducted hostelry for a town of ten thousand. The "White Hart" must be a favorite among English innkeepers, for I recollect that we stopped at no fewer than seven hotels bearing this name during our tour and saw the familiar sign on many others. On our arrival we learned that the Dukeries trip must be made by carriage and that the fifty miles would consume two days. We felt averse to subtracting so much from our already short remaining time, and when we found still further that admission was denied for the time at two of the most important estates, we decided to proceed without delay. The motor would be of no advantage to us in visiting the Dukeries, for the circuit must be made in a staid and leisurely English victoria.

Since this chronicle was written, however, I have learned that the embargo on motoring through the Dukeries is at least partially raised—another step showing the trend in England in favor of the motor car. By prearrangement with the stewards of the various estates, permission may be obtained to take a car through the main private roads. Thus the tourist will be enabled in half a day to accomplish what has previously required at least two days driving with horse and carriage.

In this vicinity is Newstead Abbey, the ancestral home of Byron, and one of our greatest disappointments was our inability to gain access to it. Perhaps we might have done so if we had made arrangements sufficiently in advance, since visitors are admitted, they told us, on certain days by special permission. There has, however, been an increasing tendency on the part of the owner to greatly limit the number of visitors. The coal mines discovered on the lands have become a great source of wealth and the abbey has been transformed into a modern palace in one of the finest private parks in England. The rooms occupied by Byron, it is said, are kept exactly as they were when he finally left Newstead and there are many interesting relics of the poet carefully preserved by the present proprietor.

It would be a bad thing for England if the tendency on the part of private owners of historic places, to exclude visitors from their premises, should become general. The disposition seems somewhat on the increase, and not without cause. Indeed, I was told that in a number of instances the privileges given had been greatly abused; that gardens had been stripped of their flowers and relics of various kinds carried away. This vandalism was not often charged against Americans, but rather against local English "trippers," as they are called—people who go to these places merely for a picnic or holiday. No doubt this could be overcome—it has been overcome in a number of instances, notably Warwick Castle and Knole House—by the charge of a moderate admission fee. People who are willing to pay are not generally of the class who commit acts of vandalism. That this practice is not adopted to a greater extent is doubtless due to the fact that numbers of aristocratic owners think there is something degrading in the appearance of making a commercial enterprise out of the historic places which they possess.

It is only twenty miles from Retford to Lincoln, and long before we reached the latter town we saw the towers of its great cathedral, which crowns a steep hill rising sharply from the almost level surrounding country. It is not strange that the cathedral-builders, always with an eye to the spectacular and imposing, should have fixed on this remarkable hill as a site for one of their churches. For miles from every direction the three massive towers form a landmark as they rise above the tile roofs of the town in sharp outline against the sky. To reach Lincoln we followed a broad, beautiful highway, almost level until it comes to the town, when it abruptly ascends the hill, which is so steep as to tax the average motor. The cathedral in some respects is the most remarkable and imposing in England. The distinctive feature is the great towers of equal size and height, something similar to those of Durham, though higher and more beautifully proportioned. The interior shows some of the finest Norman architecture in the Kingdom and the great Norman doorway is said to be the most perfect of its kind. Near the chapel in the cathedral close is a bronze statue of Tennyson accompanied by his favorite dog. This reminded us that we were in the vicinity of the poet's birthplace, and we determined that the next point in our pilgrimage should be Somersby, where the church and rectory of Tennyson's father still stand.

We planned to reach Boston that evening, and as there were a good many miles before us we were not able to give the time that really should be spent in Lincoln. It has many ancient landmarks, the most remarkable being a section of the Roman wall that surrounded the town about 15 A.D. and in which the arch of one of the gateways is still entire. It now appears to have been a very low gateway, but we were informed that excavations had shown that in the many centuries since it was built the earth had risen no less than eight feet in the archway and along the wall. Lincoln Castle, much decayed and ruinous, is an appropriate feature of one of the public parks. Along the streets leading up Cathedral Hill are rows of quaint houses, no doubt full of interest; but a motor tour often does not permit one to go much into detail.

SOMERSBY RECTORY, BIRTHPLACE OF TENNYSON.

So we bade farewell to Lincoln, only stopping to ask the hostler for directions to the next town on our way. Generally such directions are something like this: "Turn to the right around the next corner, pass two streets, then turn to the left, then turn to the right again and keep right along until you come to the town hall"—clock tower, or something of the kind—"and then straight away." After you inquire two or three more times and finally come to the landmark, you find three or four streets, any one of which seems quite as "straight away" as the others, and a consultation with a nearby policeman is necessary, after all, to make sure you are right. When once well into the country, the milestones, together with the finger-boards at nearly every parting of the ways, can be depended on to keep you right. These conveniences, however, are by no means evenly distributed and in some sections a careful study of the map and road-book is necessary to keep from going astray.

The twenty miles to Somersby went by without special incident. This quaint little hamlet—it can hardly be called a village—is almost hidden among the hills, well off the main-traveled roads and railway. We dashed through the narrow lanes, shaded in many places by great over-arching trees and the road finally led across the clear little brook made famous by Tennyson's verse. After crossing the bridge we were in Somersby—if such an expression is allowable. Nothing is there except the rectory, the church just across the way, the grange, and half a dozen thatched cottages. A discouraging notice in front of the Tennyson house stated positively that the place would not be shown under any conditions except on a certain hour of a certain day of the week—which was by no means the day nor the hour of our arrival. A party of English teachers came toward us, having just met with a refusal, but one of them said that Americans might have an exception made in their favor. Anyway, it was worth trying.

Our efforts proved successful and a neat, courteous young woman showed us over the rambling house. It is quite large—and had to be, in fact, to accommodate the rector's family of no fewer than twelve children, of whom the poet was the fourth. The oddest feature is the large dining room, which has an arched roof and narrow, stained-glass windows, and the ceiling is broken by several black-oak arches. At the base of each of these is a queer little face carved in stone and the mantel is curiously carved in black oak—all of this being the work of the elder Tennyson himself. There is some dispute as to the poet's birthroom. Our fair guide showed us all the rooms and said we might take our choice. We liked the one which opened on the old-fashioned garden at the rear of the house, for as is often the case in England, the garden side was more attractive than the front. Just across the road stands the tiny church of which the Rev. Tennyson was rector for many years. This was one of the very smallest that we visited and would hardly seat more than fifty people altogether. It is several hundred years old, and in the churchyard is a tall, Norman cross, as old as the church itself.

SOMERSBY CHURCH.

A rare thing it is to find the burying-ground around a church in England quite neglected, but the one at Somersby is the exception to the rule. The graves of the poet's father and brother were overgrown with grass and showed evidences of long neglect. We expressed surprise at this, and the old woman who kept the key to the church replied with some bitterness that the Tennysons "were ashamed to own Somersby since they had become great folks." Anyway, it seems that the poet never visited the place after the family left in 1837. Near the church door was a box with a notice stating that the congregation was small and the people poor, and asking for contributions to be used in keeping the church in repair. The grange, near the rectory, is occupied by the squire who owns the birthplace, it is a weatherbeaten building of brick and gray stone and perhaps the "gray old grange" referred to in "In Memoriam." Altogether, Somersby is one of the quietest and most charming of places. Aside from its connection with the great poet, it would be well worthy of a visit as a bit of rural England. Scattered about are several great English elms, which were no doubt large trees during the poet's boyhood, a hundred years ago.

For a long distance our road from Somersby to Boston ran on the crest of a hill, from which we had a far-reaching view over the lovely Lincolnshire country. Shortly after, we left the hills and found ourselves again in the fen country. Many miles before we reached Boston we saw the great tower of St. Botolph's Church, in some respects the most remarkable in England. They give it the inartistic and inappropriate appellation of "The Stump," due to the fact that it rises throughout its height of more than three hundred feet without much diminution in size. So greatly does this tower dominate the old-fashioned city that one is in danger of forgetting that there is anything else in Boston, and though it is a place little frequented by Americans, there are few quainter towns in England. Several hundred years ago it was one of the important seaports, but it lost its position because the river on which it is situated is navigable only by small vessels at high tide.

Boston is of especial interest to Americans on account of its great namesake in this country and because it was the point from which the Pilgrim Fathers made their first attempt to reach America. Owing to pestilence and shipwreck, they were compelled to return, and later they sailed in the Mayflower on a more successful voyage from Plymouth. We can get a pretty good idea of the reasons which led the Pilgrim Fathers to brave everything to get away from their home land. One may still see in the old town hall of Boston the small, windowless stone cells where the Fathers were confined during the period of persecution against the Puritans. Evidently they did not lay their sufferings against the town itself, or they would hardly have given the name to the one they founded in the New World. Boston is full of ancient structures, among them Shodfriars Hall, one of the most elaborate half-timbered buildings in the Kingdom. The hotels are quite in keeping with the dilapidation and unprogressiveness of the town and there is no temptation to linger longer than necessary to get an idea of the old Boston and its traditions.

The country through which we traveled next day is level and apparently productive fanning land. The season had been unusually dry and favorable to the fen land, as this section is called. The whole country between Boston and Norwich has scarcely a hill and the numerous drains showed that it is really a reclaimed marsh.

In this section English farming appeared at its best. The crops raised in England and Scotland consist principally of wheat, oats and various kinds of grasses. Our Indian corn will not ripen and all I saw of it was a few little garden patches. The fen country faintly reminds one of Holland, lying low and dotted here and there with huge windmills. As a matter of curiosity, we visited one of the latter. The miller was a woman, and with characteristic English courtesy she made us acquainted with the mysteries of the ancient mill, which was used for grinding Indian corn for cattle-feed.

Our route for the day was a circuitous one, as there were numerous points that we wished to visit before coming to Norwich for the night. A broad, level road leads from Boston to King's Lynn, a place of considerable size. Its beginning is lost in antiquity, and a recent French writer has undertaken to prove that the first settlement of civilized man in Britain was made at this point. We entered the town through one of the gateways, which has no doubt been obstructing the main highway for several hundred years. It is a common thing in the English towns to find on the main street one of the old gates, the opening through which will admit but one vehicle at a time, often making it necessary to station a policeman on each side to see that there are no collisions. But the gateways have been standing for ages and it would be sacrilege to think of tearing them down to facilitate traffic. Just outside King's Lynn we passed Sandringham Palace, a spacious modern country house and one of the favorite homes of the Royal Family.

A few hours through winding byways brought us to the village of Burnham Thorpe, the birthplace of Admiral Nelson. It is a tiny hamlet, whose mean-looking, straggling cottages with red tiles lack the artistic beauty of the average English village—the picturesque, thatched roofs and brilliant flower gardens were entirely wanting. The admiral was the son of the village rector, but the parsonage in which he was born was pulled down many years ago. Still standing, and kept in good repair, is the church where his father preached. The lectern, as the pulpit-stand in English churches is called, was fashioned of oak taken from Nelson's flagship, the Victory. The father is buried in the churchyard and a memorial to Nelson has been erected in the church. The tomb of the admiral is in St. Paul's Cathedral in London.

From Burnham Thorpe on the way to Norwich are the scant ruins of the priory of Walsingham. In its palmy days this was one of the richest in the world, and it is said that it was visited by more pilgrims than was the shrine of Becket at Canterbury. In every instance a gift was expected from the visitor, and as a consequence the monks fared sumptuously. Among these pilgrims were many of the nobility and even kings, including Henry VIII, who, after visiting the priory as a votary in the early part of his reign, ordered its complete destruction in 1539. This order was evidently carried out, for only shattered fragments of the ruins remain to show how splendid the buildings must once have been.

ST. BOTOLPH'S CHURCH FROM THE RIVER, BOSTON.

Walsingham is an unusually quaint little village, with a wonderful, ancient town pump of prodigious height and a curious church with a tall spire bent several degrees from the perpendicular. Near the priory are two springs, styled Wishing Wells, which were believed to have miraculous power, the legend being that they sprang into existence at the command of the Virgin. This illustrates one of the queer and not unpleasing features of motoring in England. In almost every out-of-the-way village, no matter how remote or small and how seldom visited by tourists, one runs across no end of quaint landmarks and historic spots with accompanying incidents and legends. Twenty miles more through a beautiful country brought us in sight of the cathedral spire of Norwich. This city has a population of about one hundred and twenty thousand and there is a unique charm in its blending of the mediaeval and modern. It is a progressive city with large business and manufacturing interests, but these have not swept away the charm of the old-time town. The cathedral is one of the most imposing in England, being mainly of Norman architecture and surmounted by a graceful spire more than three hundred feet in height. Norwich also presents the spectacle of a modern cathedral in course of building, a thing that we did not see elsewhere in England. The Roman Catholic Church is especially strong in this section, and under the leadership of the Duke of Norfolk has undertaken to build a structure that will rival in size and splendor those of the olden time. No doubt the modern Catholics bear in mind that their ancestors built all the great English churches and cathedrals and that these were lost to them at the time of the so-called Reformation of Henry VIII. Religious toleration does not prevail to any such extent in England as in the United States and there is considerable bitterness between the various sects.

Speaking of new cathedrals, while several are being built by the Roman Catholics, only one is under construction by the Church of England—the first since the days of the Stuarts. This is at Liverpool and the foundations have barely been begun. The design for the cathedral was a competitive one selected from many submitted by the greatest architects in the world. The award was made to Gilbert Scott, a young man of only twenty-one and a grandson of the famous architect of the same name who had so much to do with the restoration of several of the cathedrals. The Liverpool church is to be the greatest in the Kingdom, even exceeding York Minster and St. Paul's in size. No attempt is made to fix the time when the building will be completed, but the work will undoubtedly occupy several generations.

In Norwich we stopped at the Maid's Head Hotel, one of the noted old-time English hostelries. It has been in business as a hotel nearly five hundred years and Queen Elizabeth was its guest while on one of her visits to the city of Norwich. Despite its antiquity, it is thoroughly up-to-date and was one of the most comfortable inns that we found anywhere. No doubt this is considerably due to a large modern addition, which has been built along the same lines as the older portion. Near the cathedral are other ancient structures among which are the two gateways, whose ruins still faintly indicate their pristine splendor of carving and intricate design. The castle, at one time a formidable fortress, has almost disappeared. "Tombland" and "Strangers' Hall" are the appellations of two of the finest half-timbered buildings that we saw. The newer portions of Norwich indicate a prosperous business town and it is supplied with an unusually good street-car system. Most of the larger English cities are badly off in this particular. York, for instance, a place of seventy-five thousand, has but one street-car line, three or four miles in length, on which antiquated horse-cars are run at irregular intervals.

XIV

PETERBOROUGH, FOTHERINGHAY, ETC

The hundred miles of road that we followed from Norwich to Peterborough has hardly the suggestion of a hill, though some of it is not up to the usual English standard. We paused midway at Dereham, whose remarkable old church is the only one we saw in England that had the bell-tower built separate from the main structure, though this same plan is followed in Chichester Cathedral. In Dereham Church is the grave of Cowper, who spent his last years in the town. The entire end of the nave is occupied by an elaborate memorial window of stained glass, depicting scenes and incidents of the poet's life and works. To the rear of the church is the open tomb of one of the Saxon princesses, and near it is a tablet reciting how this grave had been desecrated by the monks of Ely, who stole the relics and conveyed them to Ely Cathedral. Numerous miracles were claimed to have been wrought by the relics of the princess, who was famed for her piety. The supposed value of these relics was the cause of the night raid on the tomb—a practice not uncommon in the days of monkish supremacy. The bones of saint or martyr had to be guarded with pious care or they were likely to be stolen by the enterprising churchmen of some rival establishment. Shortly afterwards, it would transpire that miracles were being successfully performed by the relics in the hands of the new possessors.

Leaving the main road a detour of a few miles enabled us to visit Crowland Abbey shortly before reaching Peterborough. It is a remarkable ruin, rising out of the flat fen country, as someone has said, "like a light-house out of the sea." Its oddly shaped tower is visible for miles, and one wide arch of the nave still stands, so light and airy in its gracefulness that it seems hardly possible it is built of heavy blocks of stone. A portion of the church has been restored and is used for services, but a vast deal of work was necessary to arrest the settling of the heavy walls on their insecure foundations. The cost of the restoration must have been very great, and the people of Crowland must have something of the spirit of the old abbey builders themselves, to have financed and carried out such a work. Visitors to the church are given an opportunity to contribute to the fund—a common thing in such cases. Crowland is a gray, lonely little town in the midst of the wide fen country. The streets were literally thronged with children of all ages; no sign of race suicide in this bit of Lincolnshire. Everywhere is evidence of antiquity—there is much far older than the old abbey in Crowland. The most notable of all is the queer three-way arched stone bridge in the center of the village—a remarkable relic of Saxon times. It seems sturdy and solid despite the thousand or more years that have passed over it, and is justly counted one of the most curious antiques in the Kingdom.

It was late when we left Crowland, and before we had replaced a tire casing that, as usual, collapsed at an inopportune moment, the long English twilight had come to an end. The road to Peterborough, however, is level and straight as an arrow. The right of way was clear and all conditions gave our car opportunity to do its utmost. It was about ten o'clock when we reached the excellent station hotel in Peterborough.

Before the advent of the railroad, Peterborough, like Wells, was merely an ecclesiastical town, with little excuse for existence save its cathedral. In the last fifty years, however, the population has increased five-fold and it has become quite on important trading and manufacturing center. It is situated in the midst of the richest farm country in England and its annual wool and cattle markets are known throughout the Kingdom. The town dates from the year 870, when the first cathedral minster was built by the order of one of the British chieftains. The present magnificent structure was completed in 1237, and so far as appearance is concerned, now stands almost as it left the builder's hands. It is without tower or spire of considerable height and somewhat disappointing when viewed from the exterior. The interior is most imposing and the great church is rich in historical associations. Here is buried Catherine of Aragon, the first queen of Henry VIII, and the body of the unfortunate Queen of Scots was brought here after her execution at Fotheringhay. King James I, when he came to the throne, removed his mother's remains to Westminster Abbey, where they now rest.

Strangely enough, the builders of the cathedral did not take into consideration the yielding nature of the soil on which they reared the vast structure, and as a consequence, a few years ago the central tower of the building began to give way and cracks appeared in the vaulting and walls. Something had to be done at once, and at the cost of more than half a million dollars the tower was taken down from top to foundation, every stone being carefully marked to indicate its exact place in the walls. The foundations were carried eleven feet deeper, until they rested upon solid rock, and then each stone was replaced in its original position. Restoration is so perfect that the ordinary beholder would never know the tower had been touched. This incident gives an idea of how the cathedrals are now cared for and at what cost they are restored after ages of neglect and destruction.

Peterborough was stripped of most of its images and carvings by Cromwell's soldiers and its windows are modern and inferior. Our attention was attracted to three or four windows that looked much like the crazy-quilt work that used to be in fashion. We were informed that these were made of fragments of glass that had been discovered and patched together without any effort at design, merely to preserve them and to show the rich tones and colorings of the original windows. The most individual feature of Peterborough is the three great arches on the west, or entrance, front. These rise nearly two-thirds the height of the frontage and it is almost a hundred feet from the ground to the top of the pointed arches. The market square of Peterborough was one of the largest we had seen—another evidence of the agricultural importance of the town. Aside from the cathedral there is not much of interest, but if one could linger there is much worth seeing in the surrounding country.

A TYPICAL BYWAY.

The village of Fotheringhay is only nine miles to the west. The melancholy connection of this little hamlet with the Queen of Scots brings many visitors to it every year, although there are few relics of Mary and her lengthy imprisonment now remaining. Here we came the next morning after a short time on winding and rather hilly byways. It is an unimportant looking place, this sleepy little village where three hundred years ago Mary fell a victim to the machinations of her rival, Elizabeth. The most notable building now standing is the quaint inn where the judges of the unfortunate queen made their headquarters during her farcial trial. Of the gloomy castle, where the fair prisoner languished for nineteen long years, nothing remains except a shapeless mass of grass covered stone and traces of the old-time moat. Much of the stone was built into cottages of the surrounding country and in some of the mansions of the neighborhood may be found portions of the windows and a few of the ancient mantel pieces. The great oak staircase which Mary descended on the day of her execution, is built into an old inn at Oundle, not far away. Thus the great fortress was scattered to the four winds, but there is something more enduring than stone and mortar,—its memories linger and will remain so long as the story of English history is told. King James, by the destruction of the castle, endeavored to show fitting respect to the memory of his mother and no doubt hoped to wipe out the recollection of his friendly relations with Queen Elizabeth after she had caused the death of Mary.

The school children of Fotheringhay seemed quite familiar with its history and on the lookout for strangers who came to the place. Two or three of them quickly volunteered to conduct us to the site of the castle. There was nothing to see after we got there, but our small guides were thankful for the fee, which they no doubt had in mind from the first. Mournful and desolate indeed seemed the straggling little village where three centuries ago "a thousand witcheries lay felled at one stroke," one of the cruelest and most pitiful of the numberless tragedies which disfigure the history of England.

From Fotheringhay we returned to the York road and followed it northward for about twenty miles. We passed through Woolsthorpe, an unattractive little town whose distinction is that it was the birthplace of Sir Isaac Newton. The thatched roof farmhouse where he was born is still standing on the outskirts of the village. At Grantham, a little farther on, we stopped for lunch at the "Royal and Angel" Hotel, one of the most charming of the old-time inns. Like nearly all of these old hostelries, it has its tradition of a royal guest, having offered shelter to King Charles I when on his endless wanderings during the Parliamentary wars. It is a delightful old building, overgrown with ivy, and its diamond-paned lattice windows, set in walls of time-worn stone, give evidence to its claims to antiquity.

We had paused in Grantham on our way to Belvoir Castle, about six miles away, the seat of the Duke of Rutland. This is one of the finest as well as most strikingly situated of the great baronial residences in England. Standing on a gently rising hill, its many towers and battlements looking over the forests surrounding it, this vast pile more nearly fulfilled our ideas of feudal magnificence than any other we saw. It is famous for its picture gallery, which contains many priceless originals by Gainsborough, Reynolds and others. It has always been open to visitors every week-day, but it chanced at the time that the old duke was dangerously ill—so ill, in fact, that his death occurred a little later on—and visitors were not admitted. We were able to take the car through the great park, which affords a splendid view of the exterior of the castle.

Near by is the village of Bottisford, whose remarkable church has been the burial place of the Manners family for five hundred years and contains some of the most complete monumental effigies in England. These escaped the wrath of the Cromwellians, for the Earl of Manners was an adherent of the Protector. In the market square at Bottisford stand the old whipping-post and stocks, curious relics of the days when these instruments were a common means of satisfying justice—or what was then considered justice. They were made of solid oak timbers and had withstood the sun and rain of two or three hundred years without showing much sign of decay. Although the whipping-post and stocks used to be common things in English towns, we saw them preserved only at Bottisford.

On leaving Bottisford, our car dashed through the clear waters of a little river which runs through the town and which no doubt gave it the name. We found several instances where no attempt had been made to bridge the streams, which were still forded as in primitive times. In a short time we reached Newark, where we planned to stop for the night—but it turned out otherwise. We paused at the hotel which the guide-book honored with the distinction of being the best in the town and a courteous policeman of whom we inquired confirmed the statement. We were offered our choice of several dingy rooms, but a glance at the time-worn furnishings and unattractive beds convinced us that if this were Newark's best hotel we did not care to spend the night in Newark. To the profound disgust of the landlady—nearly all hotels in England are managed by women—we took our car from the garage and sought more congenial quarters, leaving, I fear, anything but a pleasant impression behind us. We paused a few minutes at the castle, which is the principal object of antiquity in Newark. It often figured in early history; King John died here—the best thing he ever did—and it sustained many sieges until it was finally destroyed by the Parliamentarians—pretty effectively destroyed, for there is little remaining except the walls fronting immediately on the river.

Though it was quite late, we decided to go on to Nottingham, about twenty miles farther, where we could be sure of good accommodation. It seemed easy to reach the city before dark, but one can hardly travel on schedule with a motor car—at least so long as pneumatic tires are used. An obstinate case of tire trouble just as we got outside of Newark meant a delay of an hour or more, and it was after sunset before we were again started on our journey. There is a cathedral at South-well, and as we permitted no cathedral to escape us, we paused there for a short time. It is a great country church of very unusual architecture, elevated to the head of a diocese in 1888. The town of Southwell is a retired place of evident antiquity and will be remembered as having been the home of Lord Byron and his mother for some time during his youth. The route which we followed to Nottingham was well off the main highway—a succession of sharp turns and steep little hills that made us take rather long chances in our flight around some of the corners. But, luckily, the way was clear and we came into Nottingham without mishap, though it became so dark that we were forced to light our lamps—a thing that was necessary only two or three times during our summer's tour.

Our route south from Nottingham was over a splendid and nearly level road that passes through Leicester, one of the most up-to-date business towns in the Kingdom. I do not remember any place outside of London where streets were more congested with all kinds of traffic. The town is of great antiquity, but its landmarks have been largely wiped out by the modern progress it has made. We did not pause here, but directed our way to Lutterworth, a few miles farther, where the great reformer, John Wyclif, made his home, the famous theologian who translated the bible into English and printed it two hundred years before the time of Martin Luther. This act, together with his fearless preaching, brought him into great disfavor with the church, but owing to the protection of Edward III, who was especially friendly to him, he was able to complete his work in spite of fierce opposition. Strangely enough, considering the spirit of his time, Wyclif withstood the efforts of his enemies, lived to a good old age, and died a natural death. Twenty years afterward the Roman Church again came into power and the remains of the reformer were exhumed and burned in the public square of Lutterworth. To still further cover his memory with obloquy, the ashes were thrown into the clear, still, little river that we crossed on leaving the town. But his enemies found it too late to overthrow the work he had begun. His church, a large, massive building with a great, square-topped tower, stands today much as it did when he used to occupy the pulpit, which is the identical one from which he preached. A bas-relief in white marble by the American sculptor, Story, commemorating the work of Wyclif, has been placed in the church at a cost of more than ten thousand dollars, and just outside a tall granite obelisk has been erected in his honor. In cleaning the walls recently, it was discovered that under several coats of paint there were some remarkable frescoes which, being slowly uncovered, were found to represent scenes in the life of the great preacher himself.

Leaving Lutterworth, we planned to reach Cambridge for the night. On the way we passed through Northampton, a city of one hundred thousand and a manufacturing place of importance. It is known in history as having been the seat of Parliament in the earlier days. A detour of a few miles from the main road leaving Northampton brought us to Olney, which for twenty years was the home of William Cowper. His house is still standing and has been turned into a museum of relics of the poet, such as rare editions of his books and original manuscripts. The town is a quiet, sleepy-looking place, situated among the Buckinghamshire hills. It is still known as a literary center and a number of more or less noted English authors live there at the present time.

Bedford, only a few miles farther on the Cambridge road, was one of the best-appearing English towns of the size we had seen anywhere—with handsome residences and fine business buildings. It is more on the plan of American towns, for its buildings are not ranged along a single street as is the rule in England. It is best known from its connection with the immortal dreamer, John Bunyan, whose memory it now delights to honor. Far different was it in his lifetime, for he was confined for many years in Bedford Jail and it was during this imprisonment that he wrote his "Pilgrim's Progress." At Elstow, a mile from Bedford, we saw his cottage, a mean-looking little hut with only two rooms. The tenants were glad to admit visitors as probable customers for postcards and photographs. The bare monotony of the place was relieved not a little by the flowers which crowded closely around it.

JOHN WYCLIF'S CHURCH, LUTTERWORTH.

Cambridge is about twenty miles from Bedford, and we did not reach it until after dark. It was Week-End holiday, and we found the main street packed with pedestrians, through whom we had to carefully thread our way for a considerable distance before we came to the University Arms. We found this hotel one of the most comfortable and best kept of those whose hospitality we enjoyed during our tour.

Cambridge is distinctly a university town. One who has visited Oxford and gone the rounds will hardly care to make a like tour of Cambridge unless he is especially interested in English college affairs. It does not equal Oxford, either in importance of colleges or number of students. It is a beautiful place, lying on a river with long stretches of still water where the students practice rowing and where the famous boat races are held.

Cambridge is rich in traditions, as any university might be that numbered Oliver Cromwell among its students. Its present atmosphere and influences, as well as those of Oxford, are vastly different from those of the average American school of similar rank; nor do I think that the practical results attained are comparable to those of our own colleges. The Rhodes scholarship, so eagerly sought after in America, is not, in my estimation, of the value that many are inclined to put upon it. Aside from the fact that caste relegates the winners almost to the level of charity students—and they told us in Oxford that this is literally true—it seems to me that the most serious result may be that the student is likely to get out of touch with American institutions and American ways of doing things.

XV

THE CROMWELL COUNTRY. COLCHESTER.

A distinguished observer, Prof. Goldwin Smith, expressed it forcibly when he said that the epitaph of nearly every ruined castle in Britain might be written, "Destroyed by Cromwell." It takes a tour such as ours to gain something of a correct conception of the gigantic figure of Oliver Cromwell in English history. The magnitude and the far-reaching results of his work are coming to be more and more appreciated by the English people. For a time he was considered a traitor and regicide, but with increasing enlightenment and toleration, his real work for human liberty is being recognized by the great majority of his countrymen. It was only as far back as 1890 that Parliament voted down a proposition to place a statue of Cromwell on the grounds of the House of Commons; but two years later sentiment had advanced so much that justice was done to the memory of the great Protector and a colossal bronze figure was authorized and erected. I know of no more impressive sight in all England than this great statue, standing in solitary grandeur near the Houses of Parliament, representing Cromwell with sword and bible, and with an enormous lion crouching at his feet. It divides honor with no other monument in its vicinity and it seems to stand as a warning to kingcraft that it must observe well defined limitations if it continues in Britain. I saw several other statues of Cromwell, notably at Manchester, Warrington and at St. Ives.

An incident illustrating the sentiment with which the Protector is now regarded by the common people came under my own observation. With a number of other sightseers, we were visiting Warwick Castle and were being shown some of the portraits and relics relating to Cromwell, when the question was raised by someone in the party as to his position in English history. A young fellow, apparently an aspirant for church honors, expressed the opinion that Cromwell was a traitor and the murderer of his king. He was promptly taken to task by the old soldier who was acting as our guide through the castle. He said, "Sir, I can not agree with you. I think we are all better off today that there was such a man as Cromwell."

That appears to be the general sentiment of the people of Great Britain, and the feeling is rapidly growing that he was distinctly the defender of the people's rights. True, he destroyed many of the historic castles, but such destruction was a military necessity. These fortresses, almost without exception, were held by supporters of King Charles, who used them as bases of operation against the Parliamentary Army. If not destroyed when captured, they were re-occupied by the Royalists and the work had to be done over again. Therefore Cromwell wisely dismantled the strongholds when they came into his possession, and generally he did his work so well that restoration was not possible, even after the Royalists regained power. The few splendid examples which escaped his wrath—notably Warwick Castle—fortunately happened at the time to be in possession of adherents of Parliament. The damage Cromwell inflicted upon the churches was usually limited to destruction of stone images, tombs and altars, as savoring of idolatry. This spirit even extended to the destruction of priceless stained-glass windows, the loss of which we can not too greatly deplore, especially since the very art of making this beautiful glass seems to be a lost one.

At Cambridge we were within easy reach of the scenes of the Protector's early life. He was born in 1599 at Huntingdon, sixteen miles distant, and was twenty years a citizen of St. Ives, only a few miles away. He was a student at Cambridge and for several years was a farmer near Ely, being a tenant on the cathedral lands. As Ely is only fifteen miles north of Cambridge, it occurred to us to attend services at the cathedral there on Sunday morning. We followed a splendid road leading through a beautiful country, rich with fields of grain almost ready for harvest.

The cathedral is one of the largest and most remarkable in England, being altogether different in architecture from any other in the Kingdom. Instead of a spire, it has a huge, castellated, octagonal tower, and while it was several hundred years in building, a harmonious design was maintained throughout, although it exhibits in some degree almost every style of church architecture known in England. Ely is an inconsequential town of about seven thousand inhabitants and dominated from every point of view by the huge bulk of the cathedral. Only a portion of the space inside the vast building was occupied by seats, and though the great church would hold many thousands of people if filled to its capacity, the congregation was below the average that might be found in the leading churches of an American town the size of Ely. One of the cathedral officials with whom I had a short talk said that the congregations averaged small indeed and were growing smaller right along. The outlook for Ely he did not consider good, a movement being on foot to cut another diocese from the territory and to make a cathedral, probably of the great church, at Bury St. Edmunds. In recent years this policy of creating new dioceses has been in considerable vogue in England, and of course is distasteful to the sections immediately affected. The services in Ely Cathedral were simpler than usual and were through well before noon.

Before returning to Cambridge we visited St. Ives and Huntingdon, both of which were closely associated with the life of Cromwell. The former is a place of considerable antiquity, although the present town may be said to date from 1689, at which time it was rebuilt after being totally destroyed by fire. One building escaped, a quaint stone structure erected in the center of the stone bridge crossing the River Ouse and supposed to have been used as a chapel by the early monks. Cromwell's connection with St. Ives began in 1628, after he had been elected to Parliament. He moved here after the dissolution of that body and spent several years as a farmer. The house which he occupied has disappeared and few relics remain of his residence in the town. In the market square is a bronze statue of the Protector, with an inscription to the effect that he was a citizen of St. Ives for several years. A few miles farther on is Huntingdon, his birthplace. It is a considerably larger town, but none of the buildings now standing has any connection with the life of the Protector. Doubtless the citizens of Huntingdon now recognize that the manor house where Cromwell was born, which was pulled down a hundred years ago, would be a valuable asset to the town were it still standing.

From Huntingdon we returned to Cambridge, having completed a circular tour of about sixty miles. We still had plenty of time to drive about Cambridge and to view from the outside the colleges and other places of interest. The streets are laid out in an irregular manner, and although it is not a large city—only forty thousand—we had considerable difficulty in finding our way back to the hotel. The University Arms is situated on the edge of a large common called "The Field." Here in the evening were several open-air religious services. One of these was conducted by the Wesleyans, or Methodists, with a large crowd at the beginning, but a Salvation Army, with several band instruments, soon attracted the greater portion of the crowd. We found these open-air services held in many towns through England and Scotland. They were always conducted by "dissenting churches"—the Church of England would consider such a proceeding as too undignified.

We wished to get an early start from Cambridge next morning, hoping to reach London that night, and accordingly made arrangements with the head waiter for an early breakfast. We told him we should probably want it at 7:30, and he looked at us in an incredulous manner. I repeated the hour, thinking he did not understand, but he said he thought at first we were surely joking. However, he would endeavor to accommodate us. If we would leave our order that evening he thought he could arrange it at the time desired, but we could easily see that it was going to upset the traditions of the staid hotel, for the breakfast hour is never earlier than nine o'clock. However, we had breakfast at 7:30 and found one other guest in the room—undoubtedly an American. He requested a newspaper and was informed that the morning papers were not received at the hotel until half past ten o'clock, although Cambridge is just fifty miles from London, or about an hour by train. The curiosity which the average American manifests to know what happened on the day previous is almost wanting in the staid and less excitable Britisher.

We were away from Cambridge by nine o'clock and soon found ourselves in a country quite different in appearance from any we had yet passed through. Our route led through Essex to Colchester on the coast. We passed through several ancient towns, the first of them being Haverhill, which contributed a goodly number of the Pilgrim Fathers and gave its name to the town of Haverhill in Massachusetts. It is an old, straggling place that seems to be little in harmony with the progress of the Twentieth Century.

Our route on leaving Haverhill led through narrow byways, which wind among the hills with turns so sharp that a close lookout had to be maintained. We paused at Heddingham, where there is a great church and a partly ruined Norman castle. The town is made up largely of cottages with thatched roofs, surrounded by the bright English flower gardens. It was typical of several other places which we passed on our way. I think that in no section of England did we find a greater number of picturesque churches than in Essex, and a collection of photographs of these, which was secured at Earl's Colne, we prize very highly.

Colchester is an interesting town, deserving of much longer time than we were able to stay. It derived its name from King Cole, the "merry old soul" of the familiar nursery rhyme. It is one of the oldest towns in England and was of great importance in Roman times. One of the largest collections of Roman relics in Britain is to be found in the museum of the castle. There are hundreds of specimens of coin, pottery, jewelry, statuary, etc., all of which were found in excavations within the city. The castle is one of the gloomiest and rudest in the Kingdom, and was largely built of Roman bricks. It is quadrangular in shape, with high walls from twenty to thirty feet thick surrounding a small court. About a hundred years ago it was sold to a contractor who planned to tear it down for the material, but after half completing his task he gave it up, leaving enough of the old fortress to give a good idea of what it was like.

The grim old ruin has many dark traditions of the times when "man's inhumanity to man" was the rule rather than the exception. Even the mild, nonresistant Quaker could not escape the bitterest persecution and in one of the dungeons of Colchester Castle young George Fox was immured and suffered death from neglect and starvation. This especially attracted our attention, since the story had been pathetically told by the speaker at the Sunday afternoon meeting which we attended at Jordans and which I refer to in the following chapter. While there is a certain feeling of melancholy that possesses one when he wanders through these mouldering ruins, yet he often can not help thinking that they deserved their fate.

Colchester suffered terribly in Parliamentary wars and only surrendered to Cromwell after sustaining a seventy-six day siege, many traces of which may still be seen. There are two or three ancient churches dating from Saxon times which exhibit some remarkable specimens of Saxon architecture. Parts of Colchester appeared quite modern and up-to-date, the streets being beautifully kept, and there were many handsome residences. Altogether, there is a strange combination of the very old and the modern in Colchester.

We left this highway at Chelmsford to visit the Greenstead Church near Chipping-Ongar, about twenty-two miles from London. This is one of the most curious churches in all England. It is a diminutive building, half hidden amidst the profusion of foliage, and would hardly attract attention unless one had learned of its unique construction and remarkable history. It is said to be the only church in England which is built with wooden walls, these being made from the trunks of large oak trees split down the center and roughly sharpened at each end. They are raised from the ground by a low brick foundation, and inside the spaces between the trunks are covered with pieces of wood. The rough timber frame of the roof is fastened with wooden pins. The interior of the building is quite dark, there being no windows in the wooden walls, and the light comes in from a dormer window in the roof. This church was built in the year 1010 to mark the resting place of St. Edmund the Martyr, whose remains were being carried from Bury to London. The town of Ongar, near by, once had an extensive castle, of which little remains, and in the chancel of the church is the grave of Oliver Cromwell's favorite daughter. A house in High Street was for some time the residence of David Livingstone, the great African explorer.

From Chipping-Ongar we followed for the third time the delightful road leading to London, passing through the village of Chigwell, of which I have spoken at length elsewhere. On coming into London, we found the streets in a condition of chaos, owing to repairs in the pavement. The direct road was quite impassable and we were compelled to get into the city through by-streets—not an easy task. In London the streets do not run parallel as in many of our American cities. No end of inquiry was necessary to get over the ten miles after we were in the city before we reached our hotel. It was not very convenient to make inquiries, either, when driving in streets crowded to the limit where our car could not halt for an instant without stopping the entire procession. We would often get into a pocket behind a slow-moving truck or street car and be compelled to crawl along for several blocks at the slowest speed.

It was just sunset when we stopped in front of the Hotel Russell. We had been absent on our tour six weeks to a day and our odometer registered exactly 3070 miles. As there were five or six days of the time that we did not travel, we had averaged about six hundred miles a week during the tour. The weather had been unusually fine for England; we had perhaps half a dozen rainy days, but only once did it rain heavily. We had now traveled a total of 4100 miles and had visited the main points of interest in the Kingdom excepting those in the country south of the city, where we planned a short tour before sailing. We remained in London a week before starting on this trip, but during that time I did not take the car out of the garage. I had come to the conclusion that outside of Sundays and holidays the nervous strain of attempting to drive an automobile in the streets of London was such as to make the effort not worth while.

BYRON'S ELM IN CHURCH YARD, HARROW.

XVI

THE HAUNTS OF MILTON AND PENN

Leaving London by the Harrow road, in course of an hour we came to the famous college town, which lies about fifteen miles north of the city. It is known chiefly for its boys' school, which was founded early in the reign of Queen Elizabeth and at which many great Englishmen received their early education. The school is situated on the top of a hill, one of the most commanding positions in the vicinity of London, and on the very summit is the Norman church. The view from this churchyard is one of the finest in England. For many miles the fertile valley of the Thames spreads out like a great park, exhibiting the most pleasing characteristics of an English landscape. On one side the descent is almost precipitous, and at the edge, in the churchyard, stands a gigantic elm—now in the late stages of decay—which still bears the sobriquet of "Byron's Elm." It is said that Byron, during his days at Harrow, would sit here for hours at a time and contemplate the beautiful scene which spread out before him. A descendant of one of the poet's friends has placed near the spot a brass tablet, inscribed with the somewhat stilted lines, On a Distant view From Harrow Churchyard,

> "Spot of my youth, whose hoary branches sigh,
> Swept by the breeze that fans the cloudless sky;
> O! as I trace again thy winding hill,
> Mine eyes admire, my heart adores thee still.
> Thou drooping elm! Beneath whose boughs I lay,
> And frequent mused the twilight hours away;
> How do thy branches, moaning to the blast,
> Invite this bosom to recall the past,
> And seem to whisper, as they gently swell,
> 'Take, while thou canst, a lingering, last farewell'"

We reached Harrow too late to attend church as we had hoped, the morning services just closing as we entered the churchyard. We saw everywhere numbers of students in Sunday garb, and an odd appearance these boys of from fifteen to eighteen presented in a costume very nearly the counterpart of an ordinary dress suit, usually set off by a high silk hat. Harrow is associated with the names of many men who attained high rank in English history and literature, some of whom strove in their boyhood days to anticipate immortality by carving their names on the wooden desks. Among these may still be seen the rudely cut letters of the names of Byron, Sheridan and Peele.

MILTON'S ROOM IN COTTAGE AT CHALFONT ST. GILES.

The town, which slopes away from the top of the hill, has an up-to-date appearance and is a favorite place for suburban residences of wealthy Londoners. The road leading down the hill from the church turned sharply out of view, and just as we were beginning the descent a gentleman hastened to us and cautioned us not to undertake it. He said that numerous motors had been wrecked in the attempt. We went down by a roundabout way, but when we came to pass the hill at its foot, we found it was not nearly so steep as some we had already passed over.

Two or three hours over narrow and generally bad roads for England brought us to the village of Chalfont St. Giles, where John Milton made his residence while writing "Paradise Lost." It is a retired little place, mere lanes leading into it. The shriek of the railroad train does not disturb its quietude, the nearest station being several miles away. The village doubtless appears much as it did in Milton's time, three hundred years ago, and the cottage which he occupied stands practically unaltered. A notice posted outside stated that the cottage would not be shown on Sunday. But such announcements had little terror for us by this time, and we found no difficulty in gaining admittance to the quaint little building. It is in the Elizabethan style, with half-timber frame and sagging tile roof. The windows have small, diamond-shaped panes of leaded glass set in rude iron frames and open on a typical English flower garden. The villagers purchased the cottage by public subscription and its preservation is thus fortunately insured. The tenant acts as caretaker and apparently takes pride in keeping the place in order. The poet's room, directly on the right when entering, is rather dark, and has a low-beamed ceiling. There is a wide fireplace with the old time appliances accompanying it, and one can imagine the blind poet sitting by his fireside on winter days or enjoying the sweetness that in summertime came through the antique windows from the flower garden. Here he dictated "Paradise Lost" to his daughter, who acted as his secretary. One can not help contrasting the unsurpassed majesty and dignity of the great poem with the humble and even rude surroundings of the cottage. Milton came here in 1665 to escape the plague which was then devastating London. His eldest daughter was at that time about seventeen years of age, and there is reason to believe that she was with him during his stay in St. Giles. We were delighted with the place, for we had seen little else more typical of old-time England than this cottage, which would have been worth seeing aside from its connection with the great epic poet. In front was the garden, a blaze of bright colors, and the walls were half hidden by climbing rose-vines in full boom—for the roses in England stay much later in the summer than they do with us. The entrance to the cottage fronts on the garden. There is no door next the street, the great chimney built on the outside leaving no room for one.

We were now in the vicinity where William Penn was born and where he lies buried. We had some trouble in finding Jordans, the little meeting-house near which is the grave of the Quaker philanthropist. Many of the people of whom we inquired did not know of its existence, and after considerable wandering through the byways we learned that we were within a mile of the place. For this distance we followed a shady lane, over-arched by trees and so ill kept that it was about as rough motoring as one will find in England. Directly at the foot of a steep hill we came upon the meeting-house, nestling in a wooded valley. It had in its plain simplicity the appearance of an ordinary cottage; with the Quakers there in no such thing as a church, for they prefer to call their places of worship simply "meeting-houses." We were surprised to find a number of people about the chapel and soon learned that we had the good fortune to arrive on one of the meeting days. These meetings had for years been held annually, but during the present summer they were being held once a month. As the Friends are not numerous in this vicinity, many of the congregation had come from long distances—some from London. We learned this in conversation with a sweet-faced, quiet-mannered lady who had all the Quaker characteristics. She said that she and her husband had come from London that day, most of the way on their cycles; that they had been in Philadelphia and knew something of America. She presented us to a benevolent-looking, white-bearded man who afterwards proved to be the leader of the meeting, simply saying, "Our friends are from Iowa." The old gentleman pressed us to remain, as the meeting would begin immediately, and we were delighted to acquiesce. There were about forty people gathered in the little room, which was not more than fifteen by twenty feet in size and supplied with the plainest straight-backed benches imaginable. It was a genuine Quaker meeting. For perhaps half an hour the congregation sat in perfect silence, and finally the old gentleman who acted as leader arose and explained—largely for our benefit, I think, as we were the only strangers present—that this was the Quaker method of worship. Unless a member of the congregation felt he had something really worth saying, he waited to speak only "as the Spirit moved him." I could not help thinking that I had been in many meetings where, if this rule had been followed, everybody would have been better off. However, in the course of a few minutes he arose again and began his talk. We had attended many services in England at noted churches and cathedrals, but for genuine Christianity, true brotherly love and real inspiration, I think the half hour talk of the old Quaker was worth them all. We agreed that it was one of our most fortunate experiences.

In the churchyard we stood before the grave of William Penn, marked by the plainest kind of a small headstone and identical with the few others beside it. We expressed wonder at this, but the lady with whom we had previously talked explained that it would be inharmonious with the Quaker idea to erect a splendid monument to any man. For many years the graves had not been marked at all, but finally it was decided that it would not be inappropriate to put up plain headstones, all of the same style, to let visitors know where the great Quaker and his family rest. And very simple were the inscriptions chiseled upon the stones. All around the meeting-house is a forest of great trees, and no other building is in the immediate vicinity. One might almost have imagined himself at a Quaker service in pioneer times in America, when the meeting-houses were really as remote and secluded as this one seemed, rather than within twenty miles of the world's metropolis, in a country teeming with towns and villages.

It was about three o'clock when we left Jordans with a view of reaching Oxford, still a good many miles away, by nightfall. In this vicinity are the Burnham beeches, made known almost everywhere by the camera and the brush of the artist. A byway runs directly among the magnificent trees, which we found as imposing as the pictures had represented—sprawling old trees, many feet in circumference, but none of very great height. Near by is Stoke-Poges church, whose memory is kept alive by the "Elegy" of the poet Gray. It is one of the best known of the English country churches and is visited annually by thousands of people. The poet and his relatives are buried in the churchyard and the yew tree under which he is said to have meditated upon the theme of the immortal poem is still standing, green and thriving. The church, half covered by ivy and standing against a background of fine trees, presents a beautiful picture. In the immediate neighborhood a monument has been raised in memory of Gray—a huge bulk of stone of inartistic and unpleasing design. The most appropriate monument of the poet is the church itself, with its yew tree, which is now known wherever the English language is spoken.

Two or three miles farther on is Windsor, with its castle, the principal residence of royalty, and Eton College, its well known school for boys. This school is more exclusive and better patronized than Harrow, and I was told that it is quite a difficult problem for the average youth to enter at all. The sons of the nobility and members of the royal family are given the preference and expenses are so high as to shut out all but the wealthy. Windsor Castle is the most imposing of its kind in the world. It is situated on the Thames River, about twenty miles from London. Crowning a gently rising hill, its massive towers and battlements afford a picturesque view from almost anywhere in the surrounding country and especially from points of vantage in the park, where one can catch glimpses of the fortress through some of the avenues of magnificent trees. On a clear day, when the towers of the castle are sharply outlined against the sky and surmounted by the brightly colored royal standards, one might easily imagine himself back in the good old days of knight-errantry. Windsor is shown to visitors at any time when the royal family is not in residence. Queen Victoria and Albert, the Prince Consort, are buried in Frogmore Park, near by, but the tombs are sacredly guarded from the public. The grounds surrounding the castle are laid out in flower gardens and parks, and the forest of more than seven thousand acres is the finest in England. It is one of the royal preserves where the king occasionally goes hunting, but it really serves more the purpose of a great public park. There are many splendid drives through the forest open to everybody, the main one leading straight away from the castle gates for about four miles and terminating at an equestrian statue of George the Third, of more or less happy memory.

A broad road leads from Windsor to Oxford; it is almost straight and without hills of consequence. It is a favorite route for motorists, and at several points were stationed bicycle couriers of the Motor Union to give warning for police traps. These guards patrolled the road and carried circular badges, red on one side and white on the other. If the white side were shown to the passing motorist, the road ahead was clear; but the red was a caution for moderate speed for several miles. This system, which we found in operation in many places, is the means of saving motor drivers from numerous fines. The bicycle courier receives a fee very thankfully and no doubt this constitutes his chief source of revenue for service rendered.

DISTANT VIEW OF MAGDALEN TOWER, OXFORD.

About ten miles from Oxford we passed through Henley-on-Thames, famed for the University rowing-matches. Here the river lies in broad still stretches that afford an ideal place for the contests. The Thames is navigable for small steamboats and houseboats from London to Oxford, a distance of sixty miles, and the shores of the stream throughout afford scenes of surpassing beauty. Just at sunset the towers of Oxford loomed in the distance, and it was easy to recognize that of Magdalen College, which rises to a height of two hundred feet. Though Oxford is one of the older of the English towns, parts of it seemed as up-to-date as any we had seen, and the Randolph Hotel compared favorably with the best we found anywhere.

The time which a tourist will devote to Oxford will depend upon his point of view. To visit the forty-four colleges in detail and to give any time to each would manifestly require several days—if not weeks—and especially would this be true if one were interested to any extent in student life in the University. Manifestly, people touring England in a motor car do not belong to the class described. In order to get the most out of the trip, there is a constant necessity for moving on. By an economical use of time, one may gain a fair idea of Oxford in a few hours. This was what we had done on a previous trip and consequently we spent little time in the city on our second visit, merely remaining over night. I think the method we pursued would be the most practical for anyone who desires to reach the most interesting points of the town in the shortest time. We engaged an experienced hack-driver, who combined with his vocation the qualities of a well informed guide as well. We told him of our limited time and asked him to make the most of it by taking us about the universities, stopping at such as would give us the best idea of the schools and of university life. He did this to our satisfaction, and as we passed the various institutions his comments gave us a general idea of each. He stopped at some of the more noted colleges, where we often found guides who conducted us about the buildings and grounds. Perhaps Magdalen College is as interesting as any. Its fine quadrangular tower is one of the landmarks of the city, and they will tell you of the quaint custom that has prevailed for many centuries of celebrating May Day morning with music from the top of the tower by a choir of boys. Magdalen has its park and gardens, and Addison's Walk—a pathway extending for considerable distance between an avenue of fine trees beside a clear little river—is reputed to have been a haunt of the great essayist when a student at the University. Next to Magdalen, the most celebrated colleges are New College, Christ Church and Merton. At the first of these Cecil Rhodes was a student, and the great promoter must have had a warm feeling for the University, since his bequest has thrown open the various colleges to more than a hundred students from all parts of the world, but principally from the United States. Practically all of the students have their quarters in connection with the colleges and meals are served in public dining rooms.

Aside from its colleges, there is much else of interest in and about Oxford. The castle, of which there are scant remains, is one of the very oldest in England and has a varied and often stirring history. During the Parliamentary War, Oxford was one of the strongholds of the king and underwent many sieges from

Cromwell's army—which was responsible for the final destruction of the castle. As a seat of learning, the town dates from the time of Alfred, who was born at Wantage, only twenty miles away. Naturally, Oxford was always prominent in ecclesiastical affairs and during the reign of Mary the three bishops of the English church suffered martyrdom there. In one of the public places of the city stands a tall Gothic monument commemorating the services of these men and incidentally putting severe strictures on the "errors" of the Roman church. The language in which this latter clause is stated caused a storm of protest when the monument was erected, but it had no more effect than did the protest against the iron-clad, anti-Catholic coronation oath of the king. The Bodleian Library, located in Oxford, is the greatest in England, with the exception of the library of the British Museum.

XVII

A CHAPTER OF DIVERS PLACES
AND EXPERIENCES

Ten miles north of Oxford is Woodstock, near which is Blenheim Palace, the seat of the Dukes of Marlborough. This great estate and imposing mansion was presented by Act of Parliament to the first Duke of Marlborough in recognition of the victory which he won over the French at Blenheim. The architect who prepared the plans for the great structure was the famous Sir John Vanbrugh, who was so noted for the generally low heavy effect of his creations. While he was still alive a wit proposed a satirical epitaph in the couplet,

> "Lie heavy on him, Earth, for he
> Laid many a heavy load on thee."

So enormous was the cost of the palace and estate that the half million pounds sterling voted by parliament was not sufficient and more than sixty thousand pounds of the great Duke's private fortune went into it as well. In his fondness for state and display, he was quite the opposite of the other great national hero, the Duke of Wellington, who was satisfied with the greatest simplicity and preferred cash to expensive palaces and great estates. As a consequence, the Dukes of Marlborough have been land-poor for several generations and until recently Blenheim Palace seemed in a fair way to be added to the already long list of ruins in Britain. Something has lately been done in the way of repair and restoration, but there are many evidences of decay still apparent.

Blenheim Palace has been shorn of many of its treasures, among them the great Sunderland Library of 80,000 volumes, sold at auction some years ago. Many valuable objects of art still remain, especially family portraits by nearly every great artist from Gainsborough to Sargent, and there is much fine statuary. The tapestries, in the state rooms, illustrating the achievements of the first Duke, are especially remarkable and were made in Belgium under his directions. But from the English view-point, no doubt the original documents pertaining to the Duke are most notable; among these is the modest note which he addressed to Queen Anne from Blenheim, announcing his "famous victory."

RINGWOOD CHURCH.

The park is one of the largest in England, but it showed many evidences of neglect and slovenly care. Some of the worst looking cattle I saw in England obstructed the ornamental stone bridge that crosses the stream flowing into a large artificial lake within the park. The driveways were not kept in the perfect manner that is characteristic of the English private park. Despite these evidences of neglect, the beauty of the place was little impaired. There are some of the finest oak trees in England and down by the lake are groups of magnificent cedars through whose branches the bright water shimmered in the sunshine. As we circled about the park, the distant views of the palace well bore out its reputation of being one of the stateliest private homes in the Kingdom. Our guide pointed out the spot where once stood the manor-house of Woodstock, torn down about a hundred years ago. In this house Princess Elizabeth was held a prisoner for a time by her sister, Queen Mary, but it is best known from the story of Walter Scott, who located here the principal scenes of "Woodstock."

The town of Woodstock has a long line of traditions, but shows little evidence of modern progress. It is a quiet, old-world little place with clean streets and many fine trees. Tradition asserts that the father of English poetry, Geoffrey Chaucer, was born here and the old house, alleged to be his birthplace, still stands in Park Street. However, the poet himself declares that London was his native city and the confiding tourist is left with the necessity of balancing the poet's own assertion on this important point against that of the Woodstock guide books. In any event, Chaucer certainly lived in Woodstock—very likely in the house assigned to him today. The town was also a residence of the Saxon kings, and here are many legends of Henry II and Fair Rosamond. Perhaps its most distinguished resident, however, was Oliver Cromwell, who put up at an inn, now a private house, while his army battered down the old palace as described by Scott.

We returned from Woodstock to Oxford and from there directed our course to Wantage, the birthplace of King Alfred the Great and, I might incidentally remark, at that time the residence of a well known expatriated New York City politician. This latter distinction did not occur to us until after we had left the town, and therefore we failed to make inquiries as to how this gentleman was regarded by his fellow-citizens of Oxfordshire. In this connection, soon afterwards I saw an amusing report in the newspapers stating that a libel suit had been brought against a British magazine for having published an article in which the ex-boss was spoken of in an uncomplimentary manner. The report stated that the case had been settled, the magazine editor paying the legal costs and retracting what he had said, as well as publishing an apology for the attack. Here we have an example of the British idea of the sacredness of private character. This politician while in America was almost daily accused by the newspapers of every crime in the calendar and never thought it worth while to enter a denial. No sooner is he fairly established in England than he brings suit against a magazine whose charges appear to have been of the mildest character. One seldom sees in English newspapers the violent attacks on individuals and the severe denunciations of public men so common in American journals. If the editor forgets himself, as in the case cited, suit for libel is sure to be brought and often proves a serious thing. While this to some extent may obstruct the freedom of the press, it is nevertheless a relief to miss the disgraceful and unwarranted attacks on public men that continually fill the columns of many American newspapers.

The road from Oxford to Wantage is a splendid one, running through a beautiful country and bordered much of the way with ancient trees. Wantage is a quiet town, lying at the foot of the hills, and is chiefly noted as the birthplace of the great Saxon king. A granite statute of Alfred stands in the market square, representing the king with the charter of English liberties in one hand and a battle-ax in the other. As he was born more than a thousand years ago, there are no buildings now standing that were connected with his history. The church is probably the oldest building—a fine example of early English architecture. Near it is buried the wife of Whittington, "Lord Mayor of Londontown." Dr. Butler, the theologian and author of "The Analogy," was born in the town and this house is still to be seen.

Leaving Wantage, the road to Reading runs along the crest of the hills, and on either side from the breezy uplands, the green fields, dashed with the gold of the ripening harvest, stretched away for many miles. This was one of the few spots in England where the view was unobstructed by fences of any kind, and while the average English hedge-row is not unpleasing, the beauty of the landscape in this instance certainly did not suffer by its absence. From Kingston-on-Thames, the perfectly kept road closely follows the river. Reading has a population of about one hundred and twenty thousand and is a place of considerable business activity. Though the city has a history stretching back to ancient times, most of the evidences of antiquity have disappeared in modern progress. It was chosen as the seat of Elizabeth's parliament when the plague was devastating London. Fragments of the old abbey hall in which this parliament met still remain and the gateway was restored a few years ago. Reading offered a stout resistance to the Commonwealth and suffered severely at Cromwell's hands. Its chief industries today are biscuit making and seed farming, which give employment to ten thousand people.

From Reading, a few miles through byways brought us to Eversley, a retired village five miles from a railway station, where the church and rectory of Charles Kingsley may be seen. The church is picturesquely situated on the hillside, with an avenue of fine yew trees leading from the gate to the door. The building has been altered a good deal since Kingsley was rector, but the pulpit from which he preached is practically the same. The rectory, which is directly by the church, is a very old building, though it has been modernized on the side fronting the road. It stands in the midst of a group of Scotch firs which were great favorites with Kingsley. Their branches almost touch the earth, while their huge trunks form a strong contrast with the dense green of the foliage. Kingsley and his wife are buried in the churchyard on the side nearest the firs. The graves are marked by a simple Runic cross in white marble bearing the names, the date, and the legend, "God is Love." Eversley and its surroundings are thoroughly typical of rural England. A quieter and more retired little place could hardly be imagined. One wonders why the great novelist and preacher spent so many years of his life here. It may have been that the seclusion was not a little conducive to his successful literary labors.

Thirty miles farther over main-traveled highways brought us for a second time to Winchester. Here we stopped for the night after an unusually long run. An early start soon brought us to Southampton, which is known everywhere as a port of arrival and departure of great merchant steamers and which, aside from its commercial importance, is one of the most ancient and interesting cities in the Kingdom. The most notable relic is a portion of the Saxon wall, the part known as the "Arcade," built in a series of arches, being the most remarkable. Close by, in a little street called Blue Anchor Lane, is a house reputed to have been the palace of King John and said to be the oldest in England, although several others contest that distinction. At the head of Blue Anchor Lane is a picturesque Tudor house, once the residence of Henry VIII and his queen, Anne Boleyn. This is open to visitors and we were shown every part of the house by the tenant, who is also custodian. With all its magnificence of carved oak and wide fireplaces, it must have been a comfortless dwelling measured by more modern ideas.

Leaving the city, we crossed Southampton Water on a steam ferry which was guided by a chain stretched from bank to bank. Two or three miles to the southward lies Netley, a small village with the remains of an abbey dating from the reign of Henry I. The road to Netley followed the shore closely, but on nearing the village suddenly entered an avenue of fine trees which so effectually concealed the ruin that we stopped directly opposite the abbey to inquire its whereabouts. Leaving the car standing in the road, we spent a quarter of an hour wandering about the ruin and trying to locate the various apartments from a hand-book. The custodian here did not act as a guide, and we were left to figure out for ourselves the intricacies of nave, refectory, cloister, etc. Only the ivy-covered walls of the building are now standing, but these are in an unusual state of completeness. The chapel or church was cruciform in shape and built in the early English style. The walls of the west end have practically disappeared, but the great east window is fairly well preserved and its most remarkable feature is its two beautifully proportioned lights, the stone tracery of which remains almost intact. A legend in connection with this abbey no doubt grew out of the desire of some of the people to prevent the destruction of the beautiful building. After the abbey had been dismantled, the church was sold to a contractor, who proceeded to tear it down for the material. He was warned in a dream by the appearance of a monk not to proceed with the work, but disregarded the warning and was killed by the falling of a portion of the wall. If incidents of this kind had happened more frequently England would no doubt be richer in historic buildings.

We were preparing to leave Netley when a man in plain clothes approached us, and civilly touching his hat, inquired if I were the owner of the motor car. I confessed that I was and he stated he was an officer and regretted that he would have to report me to the police captain for leaving the car standing on a public walk. I had inadvertently left the machine so that it partially obstructed the narrow gravel walk alongside the road, and some of the citizens had no doubt complained to the officer. We were naturally enough much chagrined, not knowing how much inconvenience and delay this incident might cause. The constable took my name and the number of the car and said I could report the circumstance myself to the captain of the police. I desired him to accompany me to call on this dignitary, but he did not seem at all anxious for the job.

This is the general procedure in England. An arrest is very seldom made in a case of this kind. The officer simply takes the name and number and the motorist can call on the proper official himself. The police system is so perfect that it would be quite useless to attempt to run away, as would happen if such a system were pursued in this country. If, in the judgment of the police official, the case should come to trial, a summons is served on the offender and the date is set. This is what I feared might happen in this case, and as it was within a week of our sailing time, I could imagine that it might cause a great deal of inconvenience.

I found the police captain's office in a neatly kept public building with a flower garden in front of it. I put the case to the captain, and after he had learned all the particulars he hastened to assure me that he would waive prosecution of the offense. He said some of the people in Netley were prejudiced against motors and no doubt were annoyed by the numerous tourists who came there to visit the abbey. Thus all the difficulties I had conjured up faded away and I had a pleasant conversation with the captain, who was a thorough gentleman. He said that the motor car was detested by many people, and no doubt with reason in some cases; but it had come to stay and forbearance and common sense were needed on part of motorist and the public generally. Much of the trouble, he stated, is due to reckless motorists who disregard the rights of other people. The week previous they had considerable difficulty in his district with an American who drove his car recklessly and defied regulations, and it was such performances that were responsible for the prejudice against the motor. This incident was my only personal experience with the British police in official capacity, barring a friendly admonition or two in London when I managed to get on the right side of the road—which is literally the wrong side in Britain.

The English police, taken as a whole, is unquestionably the most efficient and best disciplined in the world. A policeman's authority is never questioned in England and his raised hand is a signal that never goes unheeded. He has neither club nor revolver and seldom has need for these weapons. He is an encyclopedia of information, and the cases where he lent us assistance both in directing us on our road and informing us as to places of interest, literally numbered hundreds. He is a believer in fair play and seldom starts out of his own accord to make anyone trouble. It is not the policeman, but the civil officials who are responsible for the police traps which in many places are conducted in a positively disreputable manner, the idea being simply to raise revenue regardless of justice and without discrimination among the offenders. Graft among British policemen is unknown and bribery altogether unheard of. Of course their task is easier than that of the average American policeman, on account of the greater prevalence of the law-abiding spirit among the people. One finds policemen everywhere. Even the country districts are carefully patrolled. The escape of a law-breaker is a difficult if not impossible thing. One seldom hears in England of a motorist running away and leaving the scene of an accident that he has caused. Another thing that greatly helps the English policeman in his work is that a captured criminal is not turned loose again as is often the case in this country. Justice is surer and swifter in England,

and as a consequence crime averages less than in most parts of the States. The murders committed yearly in Chicago outnumber many times those of London, which is three times as large. The British system of administering justice is one that in many particulars we could imitate to advantage in this country.

A SURREY LANDSCAPE.

From Painting by D. Sherrin.

After bidding farewell to my friend the police captain and assuring him I was glad that our acquaintance terminated so quickly and happily, we proceeded on our way towards Chichester. The road for a distance of twenty-five miles led through an almost constant succession of towns and was frightfully dusty. The weather was what the natives call "beastly hot," and really was as near an approach to summer as we had experienced so far.

The predominating feature of Chichester is its cathedral, which dates from about 1100. It suffered repeatedly from fires and finally underwent complete restoration, beginning in 1848. The detached bell-tower is peculiar to the cathedral. This, although the most recent part of the building, appeared to be crumbling away and was undergoing extensive repairs. The cathedral is one of lesser importance among the great English churches, though on the whole it is an imposing edifice.

At Chichester we stopped for lunch at the hotel, just opposite the cathedral, where we had an example of the increasing tendency of hotel managers to recoup their fortunes by special prices for the benefit of tourists. On entering the dining room we were confronted with large placards conveying the cheerful information that luncheon would cost five shillings, or about $1.25 each. Evidently the manageress desired the victims to be prepared for the worst. There was another party in the dining room, a woman with five or six small children, and a small riot began when she was presented with a bill of five shillings for each of them. The landlady, clad in a low-necked black dress with long sweeping train, was typical of many we saw in the old-country hotels. She received her guest's protest with the utmost hauteur, and when we left the altercation was still in progress. It was not an uncommon thing in many of the dingiest and most unpretentious hotels to find some of the women guests elaborately dressed for dinner in the regulation low neck and long train. In many cases the example was set by the manageress and her assistants, though their attire not infrequently was the worse for long and continuous use.

Directly north of Chichester lie the picturesque hills of Surrey, which have not inaptly been described as the play-ground of London. The country around Chichester is level bordering on the coast. A few miles to the north it becomes rough and broken. About twenty miles in this direction is Haselmere, with many associations of George Eliot and Tennyson. This, together with the picturesque character of the country, induced us to turn our course in that direction, although we found a number of steep hills that were as trying as any we had met with. On the way we passed through Midhurst, one of the quaintest of Surrey towns, situated on a hill so steep and broken as to be quite dangerous. Not far from this place is the home of Richard Cobden, the father of English free trade, and he is buried in the churchyard near the town. He was evidently held in high regard in his time, for his house, which is still standing, was presented him by the nation. Among the hills near the town are several stately English country houses, and about half a mile distant are the ruins of Cowdray mansion, which about a hundred years ago was one of the most pretentious of all. There was an old tradition which said that the house and family should perish by fire and water, and it was curiously enough fulfilled when the palace burned and the last lord of the family was drowned on the same day.

WINDMILL NEAR ARUNDEL, SUSSEX.

XVIII

IN SURREY AND SUSSEX

Twenty miles over a narrow road winding among the hills brought us to Shottermill, where George Eliot spent much of her time after 1871—a pleasant little hamlet clinging to a steep hillside. The main street of the village runs up the hill from a clear little unbridged stream, over whose pebbly bottom our car dashed unimpeded, throwing a spray of water to either side. At the hilltop, close to the church, is the old-fashioned, many-gabled cottage which George Eliot occupied as a tenant and where she composed her best known story, "Middlemarch." The cottage is still let from time to time, but the present tenant was away and the maid who answered us declined to show the cottage in her mistress' absence—a rather unusual exhibition of fidelity. The village, the surrounding country, and the charming exterior of the cottage, with its ivy and climbing roses, were quite enough to repay us for coming though we were denied a glimpse of the interior.

Haselmere is only a mile distant—a larger and unusually fine-looking town with a number of good hotels. It is a center for tourists who come from London to the Hindhead District—altogether one of the most frequented sections of England. The country is wild and broken, but in late summer and autumn it is ablaze with yellow gorse and purple heather and the hills are covered with the graceful Scotch firs. All about are places of more or less interest and a week could be spent in making excursions from Haselmere as a center. This country attracted Tennyson, and here he built his country seat, which he called Aldworth. George Eliot often visited him at this place. The house is surrounded by a park and the poet here enjoyed a seclusion that he could not obtain in his Isle of Wight home. Aldworth belongs to the present Lord Tennyson, son of the poet, who divides his time between it and Farringford in the Isle of Wight, and neither of the places are shown to visitors. However, a really interested party might see the house or even live in it, for we saw in the window of a real estate man in Haselmere a large photograph of Aldworth, with a placard announcing that it was to be "let furnished"—doubtless during the period of the year the owner passes at Farringford House.

Much as we wished to tarry in this vicinity, our time was so limited that we were compelled to hasten on. It was nearly dark when we reached Arundel, whose castle, the residence of the Duke of Norfolk, was the stateliest private mansion we saw in England. The old castle was almost dismantled by Cromwell's troops, but nearly a hundred years ago restoration was begun by the then Duke of Norfolk. It was carried out as nearly as possible along the lines of the old fortress, but much of the structure was rebuilt, so that it presents, as a whole, an air of newness. The great park, one of the finest in England, is open to visitors, who may walk or drive about at will. The road into the town leads through this park for many miles. Bordered on both sides by ancient trees and winding between them in graceful curves, it was one of the most beautiful that we had seen anywhere.

ARUNDEL CASTLE.

We had planned to stop at Arundel, but the promise in our guide-books of a "level and first-class" road to Brighton, and the fact that a full moon would light us, determined us to proceed. It proved a pleasant trip; the greater part of the way we ran along the ocean, which sparkled and shimmered as it presented a continual vista of golden-hued water stretching away toward the moon. It was now early in August; the English twilights were becoming shorter, and for the third time it was necessary to light the gas-lamps. We did not reach the hotel in Brighton until after ten o'clock.

Brighton is probably the most noted seaside resort in England — a counterpart of our American Atlantic City. It is fifty miles south of London, within easy reach of the metropolis, and many London business men live here, making the trip every day. The town has a modern appearance, having been built within the past hundred years, and is more regularly laid out than the average English city. For two or three miles fronting the beach there is a row of hotels, some of them most palatial. The Grand, where we stopped, was one of the handsomest we saw in England. It has an excellent garage in connection and the large number of cars showed how important this branch of hotel-keeping had become. There is no motor trip more generally favored by Londoners than the run to Brighton, as a level and nearly straight road connects the two cities. There is nothing here to detain a tourist who is chiefly interested in historic England. About a hundred years ago the fine sunny beach was "discovered" and the fishing village of Brightholme was rapidly transformed into one of the best built and most modern of the resort towns in England. Its present population of over one hundred thousand places it at the head of the exclusive watering places, so far as size is concerned.

A little to the north of Brighton is Lewes, the county town of Sussex, rich in relics of antiquity. Its early history is rather vague, but it is known to have been an important place under the Saxon kings. William the Conqueror generously presented it to one of his followers, who fortified it and built the castle the ruins of which crown the hill overlooking the town. The keep affords a vantage point for a magnificent view, extending in every direction. I had seen a good many English landscapes from similar points of vantage, notably the castles of Ludlow, Richmond, Raglan, Chepstow and others, and it seemed strange that in such a small country there should be so many varying and distinctly dissimilar prospects, yet all of them pleasing and picturesque.

The country around Lewes is hilly and rather devoid of trees. It is broken in many places by chalk bluffs, and the chalky nature of the soil was noticeable in the whiteness of the network of country roads. Many old houses are still standing in the town and one of these is pointed out as the residence of Anne of Cleves, one of the numerous wives of Henry VIII. Near the town and plainly visible from the tower is the battlefield where in 1624 the Battle of Lewes was fought between Henry VII and the barons, led by Simon de Montfort. Lewes appears to be an old, staid and unprogressive town. No doubt all the spirit of progress in the vicinity has been absorbed by the city of Brighton, less than a dozen miles away. If there has been any material improvement in Lewes for the past hundred years, it is hardly apparent to the casual observer.

PEVENSEY CASTLE, WHERE THE NORMANS LANDED.

We were now in a section of England rich in historic associations. We were nearing the spot where William the Conqueror landed and where the battle was fought which overthrew the Saxon dynasty—which an eminent authority declares to have done more to change the history of the Anglo-Saxon race than any other single event. From Lewes, over crooked, narrow and rather rough roads, we proceeded to Pevensey, where the Normans landed nearly a thousand years ago. It is one of the sleepy, unpretentious villages that dot the southern coast of England, but it has a history stretching far back of many of the more important cities of the Kingdom. It was a port of entry in early times and is known to have been in existence long before the Romans came to Britain. The Romans called it Anderida, and their city was situated on the site of the castle. Like other Sussex towns, Pevensey lost its position as a seaport owing to a remarkable natural movement of the coast line, which has been receding for centuries. When the Conqueror landed the sea came up to the castle walls, but now there is a stretch of four miles of meadowland between the coast and the town.

The castle, rude and ruinous, shows the work of many centuries, and was really a great fortress rather than a feudal residence. It has been in a state of decay for many hundreds of years, but its massive walls, though ivy-grown and crumbling, still show how strongly it was built. It is now the property of the Duke of Devonshire, who seeks to check further decay and opens it to the public without charge.

Battle, with its abbey, is a few miles from Pevensey. This abbey marks the site of the conflict between the Normans and the Saxons and was built by the Conqueror on the spot where Harold, the Saxon king, fell, slain by a Norman arrow. William had piously vowed that if he gained the victory he would commemorate it by building an abbey, and this was the origin of Battle Abbey. William took care, however, to see that it was filled with Norman monks, who were granted extraordinary privileges and treasure, mostly at the expense of the conquered Saxons. The abbey is one of the best preserved of the early monastic buildings in England, and is used as a private residence by the proprietor. The church is in ruins, but the great gateway, with its crenelated towers, and the main part of the monastic building are practically as they were when completed, shortly after the death of the Conqueror.

Battle Abbey, since the time of our visit, has passed into the possession of an American, who has taken up his residence there. This case is typical of not a few that came to our attention during our stay in England. Many of the historic places that have for generations been in the possession of members of the nobility have been sold to wealthy Americans or Englishmen who have made fortunes in business. These transactions are made possible by a law that permits entailed estates to be sold when the owner becomes embarrassed to such an extent that he can no longer maintain them. And some of these places are sold at astonishingly low figures—a fraction of their cost. It is another of the signs of the changing social conditions in the British Empire.

A quaint old village is Winchelsea, on the coast about fifteen miles from Battle. It is a small, straggling place, with nothing but its imposing though ruinous church and the massive gateways of its ancient walls remaining to indicate that at one time it was a seaport of some consequence. But here, as at Pevensey, the sea receded several miles, destroying Winchelsea's harbor. Its mosts interesting relic is the parish church, built about 1288. The greater portion of this is now in ruins, nothing remaining but the nave, which is still used for services. In the churchyard, under a great tree, still standing, John Wesley preached his last open-air sermon.

WINCHELSEA CHURCH AND ELM TREE.

Two miles from Winchelsea is Rye, another of the decayed seaports of the southeast coast. A few small fishing vessels still frequent its harbor, but the merchant ships, which used to contribute to its prosperity, are no longer seen. It is larger than Winchelsea and is built on a hill, its steep, narrow streets being lined with quaint houses. These two queer towns seem indeed like an echo from the past. It does not appear that there have been any changes of consequence in them for the past several hundred years. People continue to live in such villages because the average Englishman has a great aversion to leaving his native land. One would think that there would be emigration from such places to the splendid lands of Western Canada, but these lands are not being taken by Englishmen, although the opportunity is being widely advertised by the Canadian Government and the various transportation companies. And yet one can hardly wonder at the reluctance of the native Englishman to leave the "tight little island," with its trim beauty and proud tradition, for the wild, unsubdued countries of the West. If loyal Americans, as we can rightly claim to be, are so greatly charmed with England, dear indeed it must be to those who can call it their native land.

Winchelsea and Rye are typical of hundreds of decayed towns throughout the Kingdom, though perhaps they are more interesting from an historic standpoint than the others. Being so near the French coast, they suffered terribly in the continual French and English wars and were burned several times by the French in their descents upon the English coast. It was nearly dark when we reached Rye; we had planned to stop there, but the uninviting appearance of the hotel was a strong factor in determining us to reach Tunbridge Wells, about thirty miles away.

We saw few more beautiful landscapes than those which stretched away under the soft glow of the English twilight from the upland road leading out of Rye. We did not have much leisure to contemplate the beauty of the scene, but such a constant succession of delightful vistas as we dashed along, together with the exhilaration of the fresh sea breeze, forms a pleasing recollection that will not be easily effaced. The twilight was beginning to fade away beneath the brilliancy of the full moon when we ran into the village of Bodiam, where stands one of the most perfect of the ancient castellated mansions to be found in the Kingdom. We paused a few minutes to view it from a distance and found ourselves directly in front of a neat-looking hotel—the Castle Inn. Its inviting appearance, our desire to see the castle more closely, and the fact that Tunbridge Wells was still a good many miles away over winding roads liberally sprinkled with steep hills, led us to make Bodiam our stopping place. There are few things that we have more reason for rejoicing over, for we saw the gray walls and towers of Bodiam Castle under the enchanting influence of a full, summer moon.

The castle was built in 1385 and appears to have been intended more as a palatial residence than a feudal fortress. Its position is not a strong one for defense, being situated on a level plain rather than upon a commanding eminence, as is usually the case with fortified castles. It was built after artillery had come into use, and the futility of erecting a structure that would stand against this new engine of destruction must have been obvious. The most remarkable feature is the wide moat which surrounds the castle. In fact, this gives it the appearance of standing on an island in the middle of a small lake. The water of the moat was nearly covered by water-lilies.

The walls of the castle are wonderfully complete, every tower and turret retaining its old-time battlements. It is supposed never to have sustained an attack by armed forces and its present condition is due to neglect and decay. From our point of view, it must have been an insanitary place, standing in the low-lying fens in the midst of a pool of stagnant water, but such reflection does not detract from its beauty. I have never seen a more romantic sight than this huge, quadrangular pile, with its array of battlements and towers rising abruptly out of the dark waters of the moat. And its whole aspect, as we beheld it—softened in outline by the mellow moonlight—made a picture that savored more of enchantment than reality.

Although the hour was late, the custodian admitted us to the ruins and we passed over a narrow bridge which crossed the moat. The pathway led through a door in the great gateway, over which still hangs suspended the iron port-cullis. Inside there was a grassy court, surrounded by the walls and ruined apartments of the castle. I ascended one of the main towers by a dilapidated stone stairway and was well repaid for the effort by the glorious moonlit prospect that stretched out before me.

When we returned to the Castle Inn, we found the landlady all attention and she spared no effort to contribute to our comfort. The little inn was cleanlier and better kept than many of the more pretentious ones. Bodiam is several miles from the railroad and but few tourists visit the castle. The principal business of the hotel is to cater to parties of English trippers who make the neighborhood a resort for fishing and hunting.

An early start from Bodiam brought us to Tunbridge Wells before ten o'clock in the morning. This city, although of considerable size, is comparatively modern and has little to detain tourists. Like Harrogate and Bath, its popularity is largely due to its mineral springs. In its immediate neighborhood, however, there are many places of interest, and we determined to make a circular tour among some of these, returning to Tunbridge Wells for the night.

A few miles from Tunbridge Wells is Offham, a little, out-of-the-way village which boasts of a queer mediaeval relic, the only one of the kind remaining in the Kingdom. This is called a quintain post and stands in the center of the village green. It consists of a revolving crossbar on the top of a tall, white post. One end of the bar is flattened and pierced with small holes, while at the other a billet of wood is suspended from a chain. The pastime consisted of riding on horseback and aiming a lance at one of the holes in the broad end of the crossbar. If the aim were true, the impact would swing the club around with violence, and unless the rider were agile he was liable to be unhorsed—rough and dangerous sport, but no doubt calculated to secure dexterity with the lance on horseback. This odd relic is religiously preserved by the village and looks suspiciously new, considering the long period since such a pastime must have been practiced. However, this may be due to the fact that the tenant of an adjoining cottage is required by the terms of his lease to keep the post in good repair, a stipulation, no doubt, to which we owe its existence.

ENTRANCE FRONT BODIAM CASTLE, SUSSEX.

In Westerham, a few miles farther on, we saw the vicarage where Gen. Wolfe, the hero of Quebec, was born. His parents were tenants of this house for a short time only, and soon after his birth they moved to the imposing residence now known as Quebec House, and here Wolfe spent the first twelve years of his life. It is a fine Tudor mansion and has been little altered since the boyhood of the great warrior. Visitors are not now admitted. There are many relics of Wolfe in Westerham, and the spot where he received his first military commission is marked by a stone with an appropriate inscription. Wolfe's memory is greatly revered in England and he is looked upon as the man who saved not only Canada, but the United States as well, to the Anglo-Saxon race.

Quite as closely connected with American history as Quebec House is the home of William Pitt, near at hand. Holwood House, as it is called, is a stately, classic building, situated in a great forest-clad park. It passed out of the hands of Pitt more than a hundred years ago, and being in possession of a private owner, is no longer open to visitors.

Passing again into the hedge-bordered byways, we came to Downe, a retired village four miles from the railway station and known to fame as having been the home of Charles Darwin. Downe House, where he lived, is still standing, a beautiful old Eighteenth Century place which was considerably altered by Darwin himself. The house at present is evidently in the hands of a prosperous owner, for it was apparent that watchful care is expended upon it. But it is in no sense a show-place and the few pilgrims who come to the town must content themselves with a glimpse from the outside.

To get a view of the place, I surreptitiously stepped through the open gateway, the house itself being some distance from the road and partially concealed by the hedges and trees in front of it. It is a rather irregular, three-story building, with lattice windows surrounded by ivy and climbing roses. It stands against a background of fir trees, with a stretch of green lawn and flowers in front, and the whole place had an air of quiet beauty and repose. On the front of the house was an ancient sun-dial, and across it, in antique letters, the legend "Time will show." I do not know whether this was placed there by Darwin or not, but it is the most appropriate answer which the great scientist might have made to his hosts of critics. Time has indeed shown, and the quiet philosopher who lived in this retired village has revolutionized the thought of the civilized world.

XIX

KNOLE HOUSE AND PENSHURST

One of the greatest show-places of England is Knole House, the seat of the Sackville-Wests, near Seven-Oaks. The owner at the time of our visit was the Lord Sackville-West who was British ambassador at Washington, where he achieved notoriety by answering a decoy letter advising a supposed British-American to vote for Grover Cleveland as being especially friendly to England. The letter created a tremendous furor in the United States, and the result was the abrupt recall of the distinguished writer from his post.

No difficulty is experienced in obtaining admission to Knole House, providing one pays the price. The thousands of tourists who come annually are handled in a most businesslike manner. An admission fee of two shillings, or about fifty cents, is charged, and at numerous stands near the gateway photographs, post cards, souvenirs and guide-books galore are sold. Motor cars are allowed to drive right up to the great gateway, where they are assigned a position and supervised by an attendant, all for the sum of one shilling. However, the show is well worth the price, and the owner of the palace is entitled to no small credit for making it so readily accessible.

The house is a fine example of the baronial residences erected just after the period of fortified castles, when artillery had rendered these fortress-mansions useless as a means of defense. It surrounds three square courts and covers about five acres; it contains three hundred and sixty-five rooms and has seven great staircases, some of them very elaborate. The collection of paintings and mediaeval furniture is one of the best in England. The pictures are of untold value, one room being filled with originals by Gainsborough and Reynolds alone. Some idea of the value of these pictures may be gained from the fact that an offer of twenty thousand pounds for one of the Gainsboroughs was refused; and there are other pictures quite as valuable, not only by English masters, but by great continental artists as well.

King James I visited Knole House and preparations were made to receive him as befitted his rank. The immense stateroom was especially furnished for the occasion at a cost, it is said, of about one hundred thousand pounds. This room has never been used since and it stands today just as it did when it served its royal occupant, though the gorgeous hangings and tapestries are somewhat dingy and worn from the dust and decay of three hundred years.

PENSHURST PLACE, HOME OF THE SIDNEYS.

It took nearly two hours to go through the parts of the house that are shown, although the parties were accompanied by guides who kept them moving along. On the afternoon of our arrival there were quite a number of visitors, five motor cars and several carriages bringing them. Knole House stands in a large park, which has the finest beeches in England, and it is really more of a show-place than a family residence. The Sackville-Wests are among the richest of the nobility and have other homes which are probably more comfortable than this impressive but unhomelike palace.

Something similar to Knole House is Penshurst Place, about ten miles away, but with an atmosphere and traditions quite different from the Sackville-West mansion. This great palace, just adjacent to the village of Penshurst, was built in the Thirteenth Century, passing shortly after into the hands of the Sidney family, with whom it has remained ever since. Of the Sidneys, one only is known wherever the English language is spoken—the gallant young knight, Sir Philip, who, when still below the age of thirty, lost his life while fighting for a forlorn cause in the Netherlands. Of all the brilliant array of statesmen, soldiers and writers who graced the reign of Queen Elizabeth, none gave greater promise than did young Sidney. Nothing is more characteristic of him than the oft-told story of how, when suffering from his death-wound on the field of Zutphen, he gave to a wounded soldier by his side the cup of water brought to him with the greatest difficulty. There are few who have received a higher or a more deserved tribute than that of the poet Watson, when he mused upon

> "the perfect knight,
> The soldier, courtier, bard in one,
> Sidney, that pensive Hesper-light
> O'er Chivalry's departed Sun."

Naturally, we were interested in the ancestral home of such a man and the many historical associations which have gathered round it. It was at the close of a busy day for us when we reached Penshurst and learned that half an hour remained before the house would be closed for the day. Admission was easily gained and ample time given to inspect such parts of the house as were shown. We entered the great park through a gateway near the church where several members of the Sidney family are buried.

The palace stands in a large open space with a level lawn in front, and the five hundred years which have passed over it have dealt kindly with it. Few of the ancient places which we had seen in England were in better state of preservation. Nor was this due so much to restoration as in many cases. It had never been intended as a fortified castle and had escaped the ravages of war which destroyed so many of the strongholds. Its most striking feature is the baronial hall with its high, open-raftered roof, maintained in general appearance and furnishing much as it was five hundred years ago. It is of great size, and in early days the tables probably furnished cheer to hundreds of revelers at a time. At one end of the room is a gallery which the musicians occupied, and at the other, our attention was called to a small opening through which the lord of the establishment could secretly witness the doings in the hall. A remarkable feature is the fireplace, situated in the center of the room and without chimney of any kind, the smoke being left to find its way out through the windows or apertures in the roof, as the case might be—a striking example of the discomforts of the good old days when knighthood was in flower.

Queen Elizabeth, who was one of the greatest royal travelers of her time, made a visit to the home of her favorite, Sidney, and the drawing room which she honored as a guest is still shown, with much of the handsome furniture which was especially made for the occasion of Her Majesty's visit. On the walls are some examples of beautifully wrought needlework and satin tapestry which tradition says is the work of the queen herself and her maidens. In the picture gallery the majority of the paintings are portraits of the Sidney family.

From Penshurst we returned to Tunbridge Wells, having covered in all about one hundred miles since leaving that town—not a very long distance for a day's motoring, but we had seen more things of interest, perhaps, than on any other day of our tour. It was a fitting close to our tour, since we had determined that we would at once return to London and bid farewell to the English highways and byways. The next morning we spent a short time looking about Tunbridge Wells. This town has been known as a watering place since 1606 and has maintained great popularity ever since. Its unique feature is the promenade, known as "The Pantiles," with its row of stately lime trees in the center and its colonade in front of the shops. It is referred to in Thackeray's "Virginians," and readers of that story will recall his description of the scenes on the Pantiles in the time of the powdered wigs, silver buckles and the fearful and wonderful "hoop." Tunbridge Wells makes a splendid center for several excursions and one might well spend considerable time there. Our trip of the previous day had taken us at no time more than thirty miles from the town and had covered only a few of the most interesting places within that distance.

We were ready to leave Tunbridge Wells before noon, and it was with feelings of mingled satisfaction and regret that we turned toward London, about thirty miles away. Our long summer's pilgrimage through Britain was over. Despite our anxiety to return home, there was, after all, a sense of regret that we had left undone much that would have been well worth while. Our last day on the English country roads was a lovely one. A light rain had fallen the night before, just enough to beat down the dust and freshen the landscape. We passed through a country thickly interspersed with suburban towns. The fields had much the appearance of a well kept park, and everything conspired to make the day a pleasant recollection.

When we came into the immediate suburbs of London, I found that the knowledge I had gained on our frequent trips gave me a great advantage in getting into the city. I was able to avoid the crowded streets and to select those where traffic was lighter, thus reducing the time of reaching our hotel fully an hour. There is much difference in the traffic on the eight bridges which cross the Thames. London Bridge, which crosses near the Bank of England, is the most congested of all. There is hardly an hour when it is not a compact mass of slowly moving vehicles. The bridge by Parliament House is less crowded, but I should say that Waterloo Bridge furnishes the best route for motorists in getting across the river. It leads directly into the new boulevard known as Kingsway, which has just been completed at an expense of many millions of pounds. This is the broadest street in London and was opened by wholesale condemnation of private property. It is little used for heavy traffic and has a fine asphalted surface. It extends from the Strand to Holborn, the two principal business arteries of London. The street now presents a rather ragged appearance on account of the buildings that were torn down to make way for it. However, new structures of fine architecture are rapidly being built and Kingsway is destined to become one of the handsomest boulevards in the world.

A little after noon we reached our London hotel, having spent ten weeks in touring England, Wales and Scotland. We had not confined ourselves to the highways, but had journeyed a great part of the distance through less frequented country roads. In fact, many of the most charming places we had visited could be reached only from the byways and were not immediately accessible from railway stations. With the exception of the first two weeks, when we had rain more or less every day, we had been favored with exceptionally fine weather. During the last seven or eight weeks of our trip, only light showers had fallen and we were assured that the season had been an unusual one for England.

The matter of weather is not of great moment to the motorist in Great Britain. The roads are not affected in the least, so far as traveling is concerned, and dashing through the open air in a rain is not an unpleasant experience. A closed top for the car is rarely necessary. Plenty of waterproof coats and coverings answer the purpose very well and the open air is much pleasanter than being cooped up in a closed vehicle. Rubber tires do not slip on good macadam roads and during our tour it was necessary to use chains on the wheels only a few times.

Altogether, the experience was worth while; nor was it so expensive as many have imagined it to be. A party of three or four people with their own car, if one of them drives, can tour Britain for less than it would cost to cover the same ground, traveling first-class, by railway train. As to the comparative satisfaction derived from the two methods of touring, no comment whatever is needed. Making the trip by motor affords so many advantages and so many opportunities of seeing the country and of coming in touch with the people that there is really no other method that can in any way compare with it.

XX

SOME MIGHT-HAVE-BEENS

In closing this desultory record of a summer's motoring in Britain, I can easily see that a great deal was missed, much of which might have been included with little or no loss of time had we been well enough informed in advance. There were cases where we actually passed through places of real interest only to learn later that we had overlooked something that might well have engaged our attention. There were other points, readily accessible from our route, which we omitted because previously visited by rail; and though many of these places we should have been glad to see again, our limited time forbade. In order to get all that should be gotten out of a five-thousand-mile tour by motor car, one would have to be familiar indeed with England's history and traditions, as well as conversant with her literature. There is little opportunity for studying hand-books as one goes along. A few weeks of preparation, of well selected reading and the study of road-books and maps would make such a tour doubly valuable in saving time and in an intelligent understanding of the country and the places worth seeing. What one should have done he will know far better after the trip is over, and the main excuse for this modest record is that it may supply in popular form some data from the experience of one who has been over part of the ground, while the superb illustrations of the volume will give a far better idea of what awaits the tourist than the mere written words.

Among the places in which our time was too short is Canterbury. Another day would have given us a chance to see more of that ancient town, and a side trip of thirty miles would have taken us to Sandwich, Margate and Reculvers. We had expected to come a second time to Canterbury and to visit these three points then, but were unable to carry out our plan. Sandwich was at one time an important seaport, but lost its position from the same cause that affected so many of the south coast towns—the receding of the sea. It contains many of the richest bits of mediaeval architecture in England, and a few hours in its quaint streets would have been well repaid. Reculvers, or ancient Regulbium, was a Roman city that was destroyed by the encroachments of the sea. Here is one of the oldest and strangest of the ruined churches in England, now standing on the verge of the ocean, which still continues to advance with a prospect of ultimately wiping out the little village.

On our trip to Manchester we passed within two or three miles of Knutsford, the delightful old town selected by Mrs. Gaskell as the scene of her story, "Granford." Had we known of this at the time, a short detour would have taken us through its quaint streets.

A BIT OF OLD ENGLAND.

From Water Color by Anderson.

The Isle of Wight is immediately across the strait from Southampton, and while a motor car could be transported by steamer to traverse its fifty or sixty miles of main road, this is not very often done. It would require one or two days to visit the interesting points in the island, among which are Carisbrooke Castle, where King Charles I was confined as a prisoner; Osborne House, formerly a royal residence but presented to the nation by King Edward; and Freshwater, the home where the poet Tennyson lived for many years.

Sherborne and Tewkesbury were both only a few miles off our route, and had we planned rightly we could have visited with very little loss of time these two interesting towns with their great abbey churches, which rank in size and importance with many of the cathedrals.

Ten miles from Penzance would have brought us to Lands End—the extreme southwestern point of England, abounding in wild and beautiful ocean-shore scenery, but the story of dangerous hills deterred us, though we afterwards regretted our decision. Nor could we pass again as we did at Camelford in Cornwall within five miles of King Arthur's Tintagel without seeing this solitary and wonderfully romantic ruin, with the majestic—even awe-inspiring—scenery around it.

Perhaps the most interesting trip which we missed, but which would have required more time than we could give, was a two or three days' run through the extreme south of Wales. It is only thirty miles from Monmouth to Cardiff, a coal-mining metropolis, itself of little interest, but with many places worth visiting in its immediate vicinity. Cardiff Castle, too, is one of the best known of the Welsh ruins, and here Henry I confined his elder brother Robert for twenty years while he himself, in reality a usurper, held the English throne. Ten miles north of Cardiff is the rude and inaccessible castle of Caerphilly, which is reckoned the most extensive ruin in the Kingdom.

Following the coast road for one hundred miles, one comes to the ancient town of St. Davids, at the extreme southwestern point of Wales. Here in the Middle Ages was a city of considerable size, a great resort of pilgrims to St. David's shrine, William the Conqueror being one of these. The modern St. Davids is a mere village, and its chief attraction is its grand cathedral and the ruins of the once gorgeous episcopal palace. The cathedral, built in the Tenth Century, is curiously situated in a deep dell, and only the great tower is visible from the village.

The return trip from St. Davids would best be made over the same road to Carmarthen, then taking the road northward to Llandovery, where is located one of the ruins of what was once the greatest abbey in Southern Wales. From this point the road direct to Abergavenny is a good one and passes through much of the picturesque hill country of Wales.

From Bangor in North Wales it is about twenty miles to Holyhead, from which point the car could easily be transferred to Ireland in two or three hours. This would mean an additional two weeks to the tour, and no doubt more time could pleasantly be spent in the Emerald Isle. The roads in Ireland are far from equal to those of England or Scotland, but the scenery, especially on the coast, is even lovelier, and the points of interest quite as numerous.

The Isle of Man, in the Irish Channel, is a famous resort of motorists, and many of the speed and reliability contests have been held there. It is about the only spot in the world where no speed limit is imposed, the inhabitants of the island recognizing the financial advantage which they reap from the numerous motorists. There are about fifty or sixty miles of road in the island said to be as fine as any in the world. The island is charming and interesting, with ruins and relics dating from the time it was an independent kingdom. The two days which would have to be given it would be well spent.

No one who had not visited it before would miss the Lake District in the north of England. A former trip through this section by coach caused us to omit it from our tour, though we would gladly have seen this delightful country a second time. One could depart from the main highway from Lancaster to Carlisle at Kendall and in a single day visit most of the haunts of Ruskin, Coleridge, Wordsworth and Southey, whose names are always associated with the English lakes. Many steep hills would be encountered, but none that would present great difficulty to a moderate-powered motor. It would be much better, however, if two or three days could be given to the Lakes, and this time might also include Furness Abbey and Lanercost Priory. Volumes have been written of the English lakes, but with all the vivid pen-pictures that have been drawn one will hardly be prepared for the beauty of the reality.

The Peak District in Derbyshire we omitted for the same reason—a previous visit. At Nottingham we were within ten or fifteen miles of this section, and by following a splendid road could have reached Rowsley Station, with its quaint inn, near Chatsworth House and Haddon Hall. No one who makes any pretense of seeing England will miss either of these places. Haddon Hall is said to be the most perfect of the baronial mansion houses now to be found in England. It is situated in a wonderfully picturesque position, on a rocky bluff overlooking the River Wye. The manor was originally given by the Conqueror to Peveril of the Peak, the hero of Scott's novel. The mansion is chiefly famous for its connection with Dorothy Vernon, who married the son of the Earl of Rutland in the time of Queen Elizabeth, the property thus passing to the Rutland family, who are still the owners. The mansion is approached by a small bridge crossing the river, whence one enters under a lofty archway the main courtyard. In this beautiful quadrangle, one of the most interesting features is the chapel at the southwest corner. This is one of the oldest portions of the structure. Almost opposite is the magnificent porch and bay-window leading into the great hall. This is exactly as it was in the days of the Vernons, and its table, at which the lord of the feast sat, its huge fireplace, timber roof and minstrel gallery are quite unaltered. It has recently been announced that the Duke of Rutland will make repairs to this old place and occupy it as one of his residences, closing Belvoir Castle, his present home, on account of the great expense of maintaining it.

Four or five miles from Haddon Hall is Chatsworth House, the splendid country seat of the Duke of Devonshire. This was built over a hundred years ago and is as fine an example of the modern English mansion as Haddon Hall is of the more ancient. It is a great building in the Georgian style, rather plain from the outside, but the interior is furnished in great splendor. It is filled with objects of art presented to the family at various times, some of them representing gifts from nearly every crowned head in Europe during the last hundred years. Its galleries contain representative works of the greatest ancient and modern artists. Even more charming than the mansion itself are its gardens and grounds. Nowhere in England are these surpassed. The mansion, with its grounds, is open daily to the public without charge, and we were told that in some instances the number of visitors reaches one thousand in a single day. As I noted elsewhere, the Duke of Devonshire owns numerous other palaces and ruins, all of which are open to the public without charge—a fine example of the spirit of many of the English nobility who decline to make commercial enterprises of their historic possessions.

In this immediate vicinity is Buxton, another of the English watering places famous for mineral springs. The neighborhood is most romantic, with towering cliffs, strange caverns, leaping cataracts and wooded valleys. However, the section abounds in very steep hills, dangerous to the most powerful motor.

In Yorkshire we missed much, chiefly on account of lack of time. A single day's journey would have taken us over a fine road to Scarborough, an ancient town which has become a modern seacoast resort, and to Whitby, with one of the finest abbey ruins in the shire, as well as to numerous other interesting places between. Barnard Castle, lying just across the western boundary of Yorkshire, was only a few miles off the road from Darlington, and would have been well worth a visit. These are only a few of the many places which might be seen to advantage if one could give at least a week to Yorkshire.

From Norwich an hour or two would have taken us to Yarmouth through the series of beautiful lakes known as the Norfolk Broads. Yarmouth is an ancient town with many points of interest and at present noted principally for its fisheries.

On the road to Colchester we might easily have visited Bury St. Edmunds, and coming out of Colchester, only seven miles away is the imposing ruin of the unfinished mansion of the Marneys, which its builder hoped to make the most magnificent private residence in the Kingdom. The death of Lord Marney and his son brought the project to an end and for several hundred years this vast ruin has stood as a monument to their unfulfilled hopes.

It may seem that as Americans we were rather unpatriotic to pass within a few miles of the ancestral country of the Washingtons without visiting it, but such was the case. It is not given much space in the guide-books and it came to us only as an afterthought. It was but five or six miles from Northampton, through which we passed. In the old church at Brington is the tomb of George Washington's great-great-great-grandfather and also one of the houses which was occupied by his relatives. In the same section is Sulgrave Manor, the home of the Washingtons for several generations, which still has over its front doorway the Washington coat-of-arms. In the same vicinity and near the farmhouse where George Eliot was born is Nuneaton, a place where she spent much of her life and to which numerous references are made in her novels.

THE CALEDONIAN COAST.

From Painting by D. Sherrin.

In Scotland we also missed much, but very little that we could have reached without consuming considerably more time. A day's trip north of Edinburgh, across the Firth of Forth into Fife, would have enabled us to visit Loch Leven and its castle, where Queen Mary was held prisoner and was rescued by young Douglas, whom she afterward unfortunately married. Had we started two or three hours earlier on our trip to Abbottsford and Melrose, we could easily have reached Jedburgh and Kelso, at each of which there are interesting abbey ruins. Of course it would have been a fine thing to go to the extreme northern point of Scotland, known as John O' Groats, but this, at the rate we traveled, would have consumed two or three days. The country is not specially interesting and has few historical associations. Tourists make this trip chiefly to be able to say they have covered the Kingdom from Lands End to John O' Groats.

I have said little of the larger cities—we did not stop long in any of these. The chief delight of motoring in Britain is seeing the country and the out-of-the-way places. In the cities, where one may spend days and where the train service and other methods of transportation in the place and its suburbs are practically unlimited, one can ill afford to linger with his car in the garage much of the time. Of London I have already spoken. Liverpool, Manchester, Leeds, Bristol, Birmingham, Edinburgh and Glasgow are examples to my point. We had visited nearly all of these by rail, but in again planning a tour by car I should not stop at such places for any length of time and should avoid passing through them whenever practicable.

Of course I do not pretend in the few suggestions I have made in this chapter to have named a fraction of the points of interest that we did not visit—only the ones which appealed to me most when I had become more familiar with Britain. I only offer these few comments to show how much more might have been compassed in the space of a week or two, leaving out Ireland, John O' Groats, and the Isles of Wight and Man. One week would have given ample time for us to include the places I have enumerated. In planning a tour, individual taste must be a large element. What will please one may not appeal so strongly to another. Still, I am sure that the greater part of the route which we covered and which I have tried to outline will interest anyone who cares enough to give the time and money necessary to tour Britain.

MAP OF ENGLAND AND WALES.

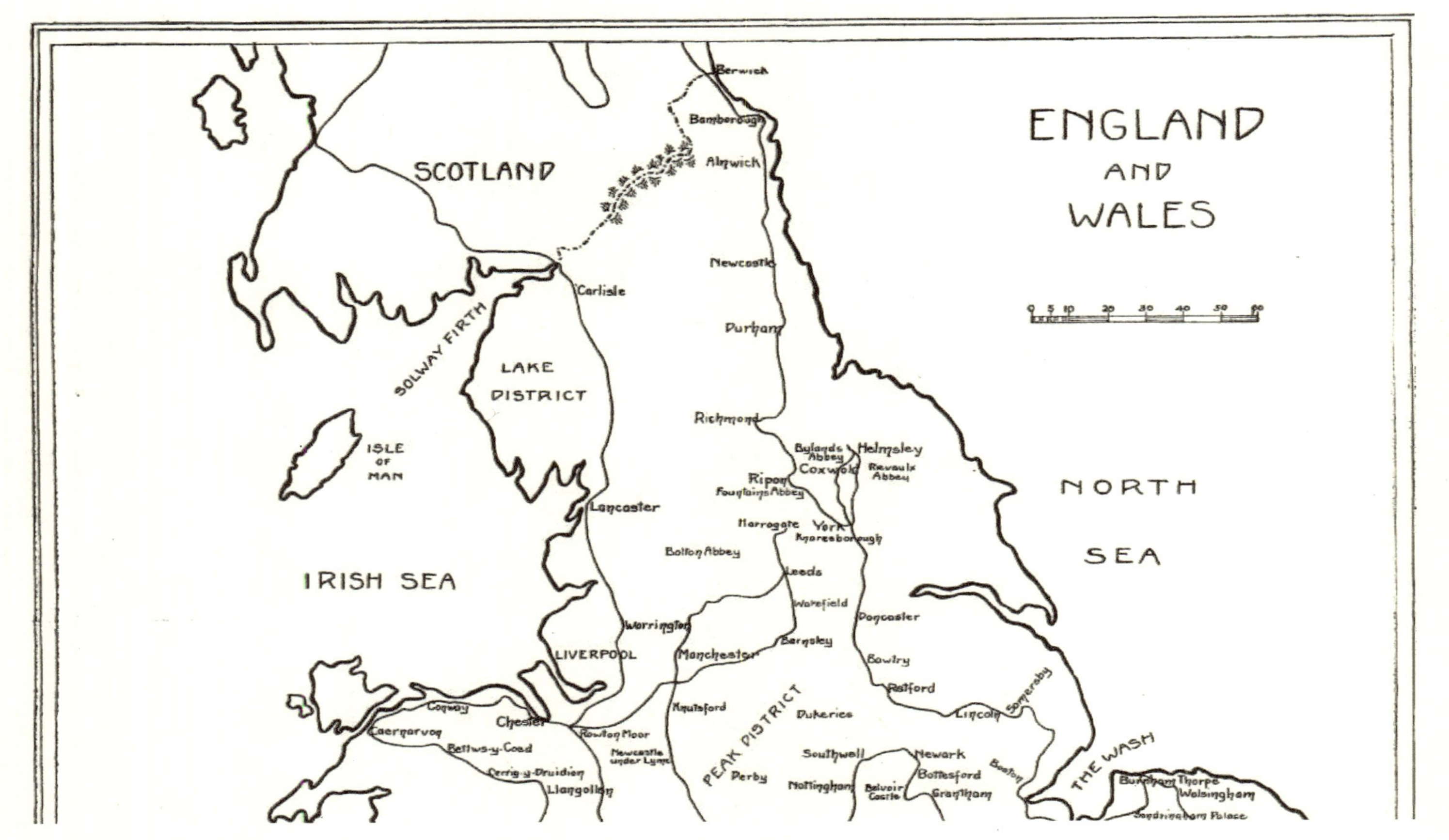

ENGLAND
AND
WALES
0 5 10 20 30 40 50 60
SCOTLAND
Berwick
Bamborough
Alnwick
Newcastle
Carlisle
Durham
SOLWAY FIRTH
LAKE DISTRICT
Richmond
Bylands Abbey
Helmsley
Coxwold
Revaulx Abbey
Ripon
Fountains Abbey
Harrogate
York
Knaresborough
ISLE OF MAN
Lancaster
Bolton Abbey
Leeds
Wakefield
Doncaster
IRISH SEA
Warrington
Barnsley
LIVERPOOL
Manchester
Bawtry
Retford
Knutsford
Dukeries
Lincoln
Sompersby
Conway
Chester
PEAK DISTRICT
Caernarvon
Rowton Moor
Southwell
Newark
THE WASH
Bettws-y-Coad
Newcastle under Lyme
Bottesford
Boston
Burnham Thorpe
Cerrig-y-Druidion
Derby
Nottingham
Belvoir Castle
Grantham
Walsingham
Llangollen
Sandringham Palace
NORTH SEA

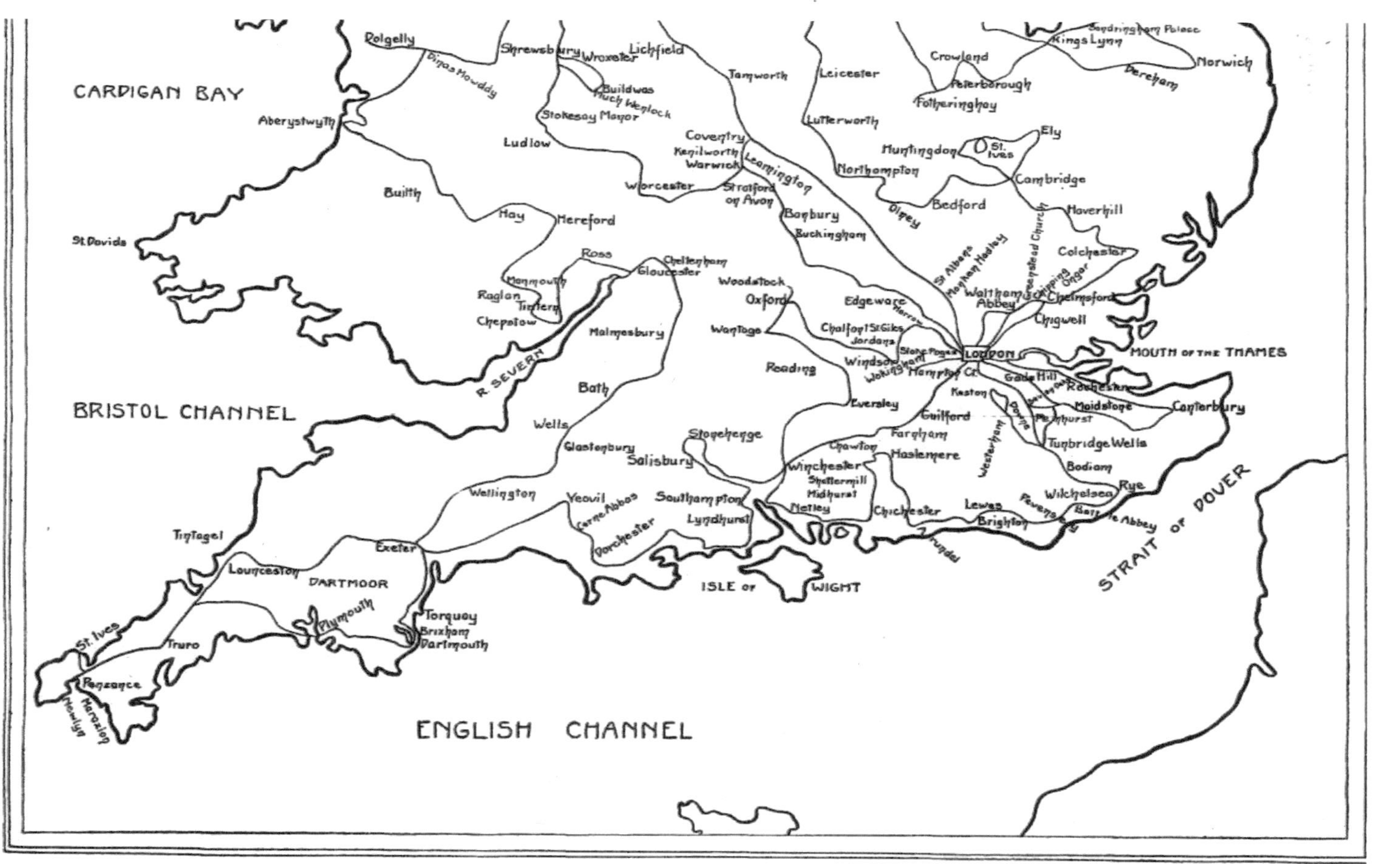

CARDIGAN BAY
BRISTOL CHANNEL
ENGLISH CHANNEL
STRAIT of DOVER
MOUTH of the THAMES
ISLE of WIGHT
DARTMOOR
R. SEVERN
Dolgelly
Dinas Mowddy
Shrewsbury
Wroxeter
Lichfield
Tamworth
Leicester
Crowland
Kings Lynn
Sandringham Palace
Dereham
Norwich
Aberystwyth
Buildwas
Much Wenlock
Stokesay Mayor
Peterborough
Fotheringhay
Ludlow
Coventry
Kenilworth
Warwick
Leamington
Lutterworth
Huntingdon
St. Ives
Ely
Builth
Worcester
Stratford on Avon
Northampton
Cambridge
Hay
Hereford
Banbury
Olney
Bedford
Haverhill
St Davids
Buckingham
Ross
Cheltenham
Gloucester
Woodstock
St Albans
Waltham Abbey
Hoddesdon
Epping
Chipping Ongar
Colchester
Monmouth
Raglan
Tintern
Chepstow
Oxford
Edgeware
Harrow
Chelmsford
Malmesbury
Wantage
Chalfont St Giles
Jordans
Chigwell
Stoke Poges
LONDON
MOUTH of the THAMES
Reading
Windsor
Eton
Woking
Hampton Ct.
Gads Hill
Rochester
Bath
Eversley
Keston
Downs
Playhurst
Maidstone
Canterbury
Wells
Guilford
Farnham
Westerham
Tunbridge Wells
Glastonbury
Salisbury
Stonehenge
Chawton
Haslemere
Bodiam
Rye
Winchester
Shottermill
Midhurst
Pevensey
Winchelsea
Wellington
Yeovil
Cerne Abbas
Southampton
Netley
Chichester
Lewes
Battle Abbey
Tintagel
Exeter
Lyndhurst
Dorchester
Brighton
Dungeness
Launceston
St Ives
Truro
Plymouth
Torquay
Brixham
Dartmouth
Penzance
Marazion
Mowlyn

MAP OF SCOTLAND.

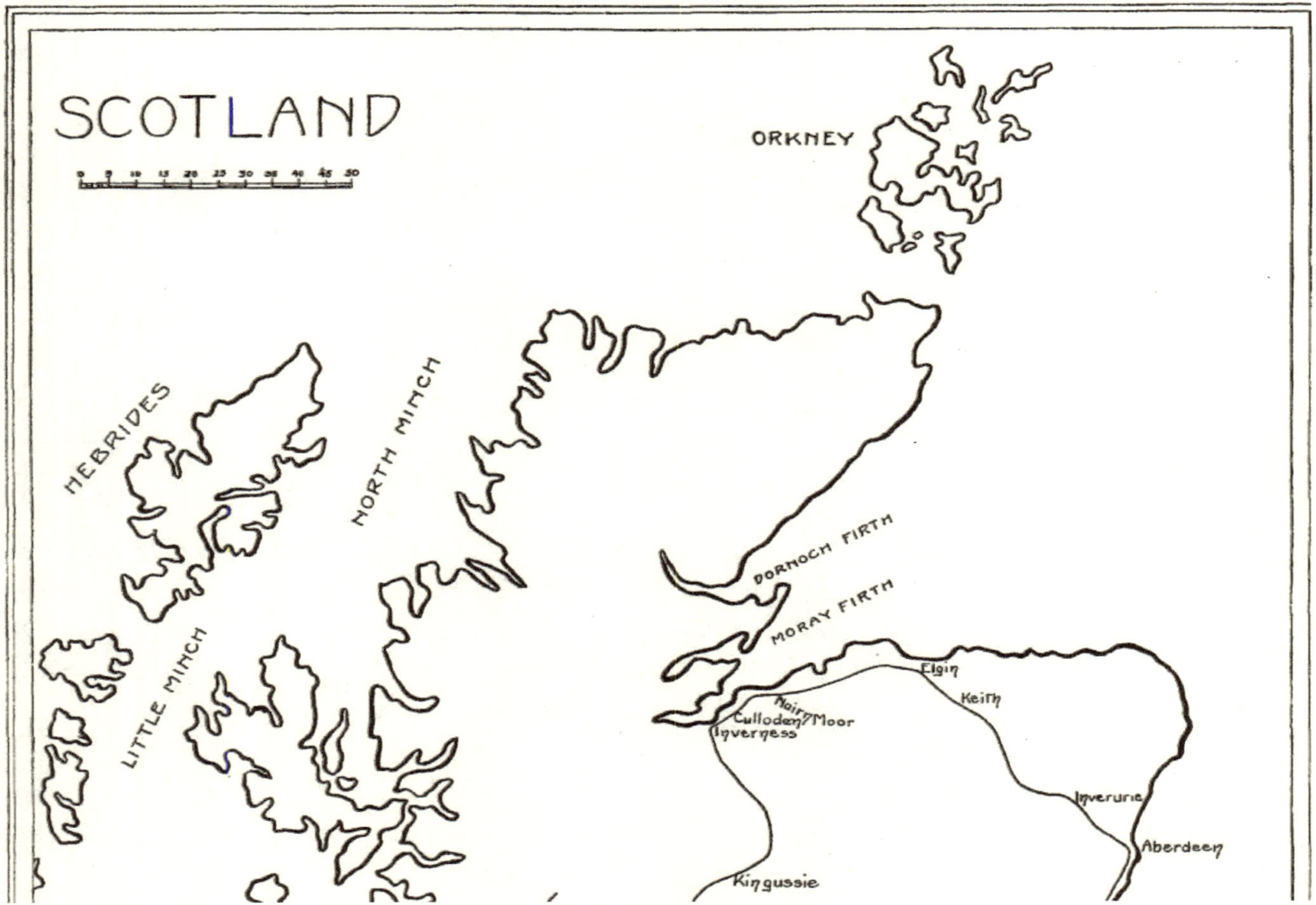

SCOTLAND
0 5 10 15 20 25 30 35 40 45 50
ORKNEY
HEBRIDES
NORTH MINCH
LITTLE MINCH
DORNOCH FIRTH
MORAY FIRTH
Elgin
Keith
Nairn
Culloden Moor
Inverness
Inverurie
Aberdeen
Kingussie

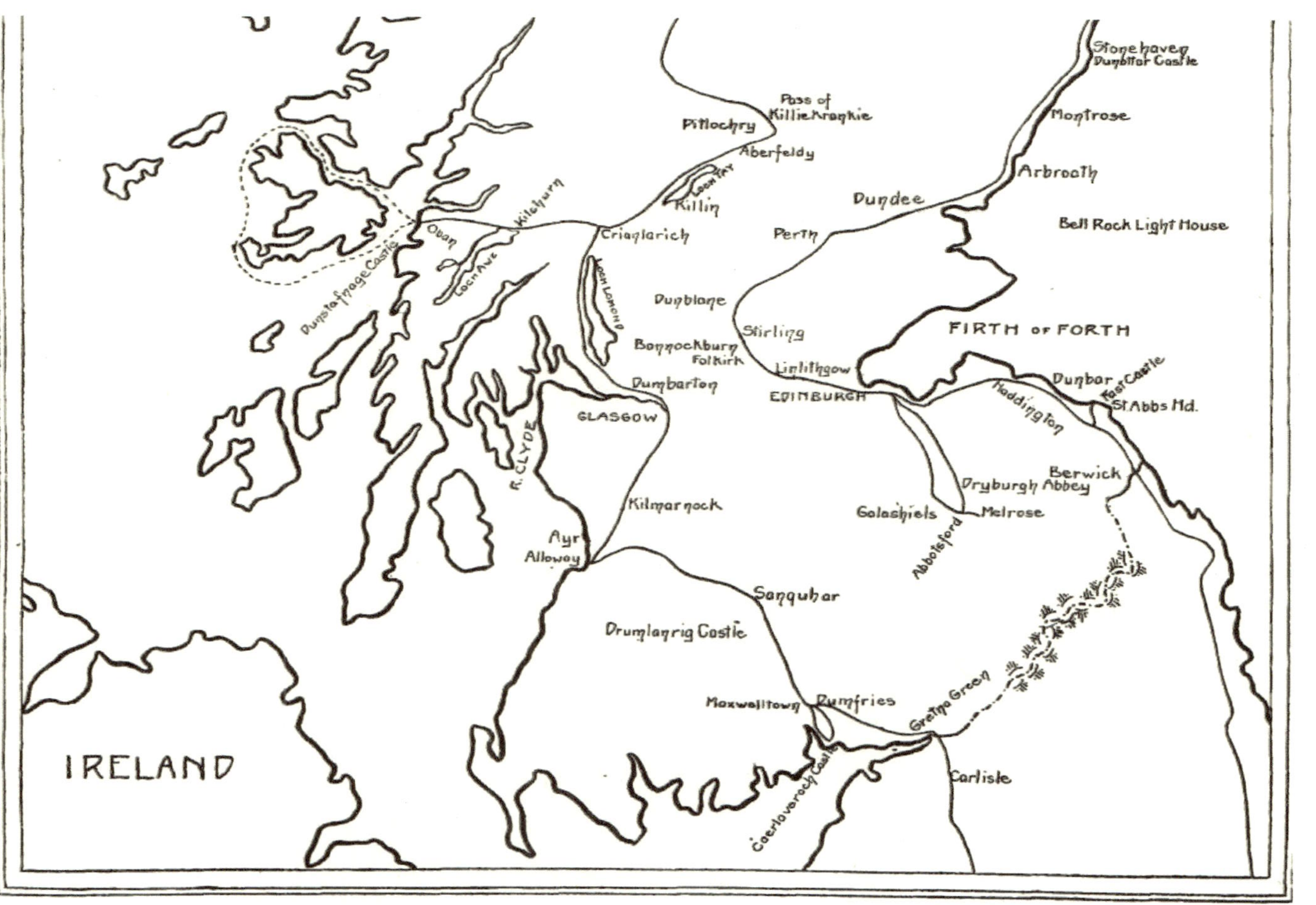

Stonehaven
Dunnottar Castle
Montrose
Arbroath
Bell Rock Light House
Pass of Killiecrankie
Pitlochry
Aberfeldy
Loch Tay
Killin
Dundee
Perth
Crianlarich
Oban
Loch Awe
Kilchurn
Dunstaffnage Castle
Loch Lomond
Dunblane
Stirling
FIRTH OF FORTH
Bannockburn
Falkirk
Linlithgow
Dumbarton
EDINBURGH
Dunbar
Fast Castle
St Abbs Hd.
Haddington
GLASGOW
R. CLYDE
Berwick
Dryburgh Abbey
Kilmarnock
Golashiels
Melrose
Abbotsford
Ayr
Alloway
Sanquhar
Drumlanrig Castle
Maxwelltown
Dumfries
Gretna Green
Carlisle
Caerlaverock Castle
IRELAND

INDEX

Lector House believes that a society develops through a two-fold approach of continuous learning and adaptation, which is derived from the study of classic literary works spread across the historic timeline of literature records. Therefore, we aim at reviving, repairing and redeveloping all those inaccessible or damaged but historically as well as culturally important literature across subjects so that the future generations may have an opportunity to study and learn from past works to embark upon a journey of creating a better future.

This book is a result of an effort made by Lector House towards making a contribution to the preservation and repair of original ancient works which might hold historical significance to the approach of continuous learning across subjects.

HAPPY READING & LEARNING!

LECTOR HOUSE
LECTOR HOUSE LLP
E-MAIL: lectorpublishing@gmail.com